SpringerBriefs in Mathematical Physics

Volume 47

SpringerBriefs are characterized in general by their size (50–125 pages) and fast production time (2–3 months compared to 6 months for a monograph).

Briefs are available in print but are intended as a primarily electronic publication to be included in Springer's e-book package.

Typical works might include:

- An extended survey of a field
- A link between new research papers published in journal articles
- A presentation of core concepts that doctoral students must understand in order to make independent contributions
- Lecture notes making a specialist topic accessible for non-specialist readers.

SpringerBriefs in Mathematical Physics showcase, in a compact format, topics of current relevance in the field of mathematical physics. Published titles will encompass all areas of theoretical and mathematical physics. This series is intended for mathematicians, physicists, and other scientists, as well as doctoral students in related areas.

Editorial Board

- Nathanaël Berestycki (University of Cambridge, UK)
- Mihalis Dafermos (University of Cambridge, UK / Princeton University, US)
- Atsuo Kuniba (University of Tokyo, Japan)
- Matilde Marcolli (CALTECH, US)
- Bruno Nachtergaele (UC Davis, US)
- Hal Tasaki (Gakushuin University, Japan)

- 50 – 125 published pages, including all tables, figures, and references
- Softcover binding
- Copyright to remain in author's name
- Versions in print, eBook, and MyCopy

Makoto Katori

Elliptic Extensions
in Statistical and Stochastic
Systems

 Springer

Makoto Katori
Department of Physics
Chuo University
Tokyo, Japan

ISSN 2197-1757   ISSN 2197-1765 (electronic)
SpringerBriefs in Mathematical Physics
ISBN 978-981-19-9526-2   ISBN 978-981-19-9527-9 (eBook)
https://doi.org/10.1007/978-981-19-9527-9

This Springer imprint is published by the registered company Springer Nature Singapore Pte Ltd.
The registered company address is: 152 Beach Road, #21-01/04 Gateway East, Singapore 189721,
Singapore

To Hiroko, Machiko, and Rieko

Preface

It is known from Hermite's theorem that there are three levels of mathematical frames in which a simple addition formula (the Riemann–Weierstrass addition formula) is valid [78, Chapter XX, Miscellaneous Examples 38] [30]. They are rational, q-analogue, and elliptic-analogue.[1] Based on the addition formula and associated mathematical structures, fruitful developments of study have been achieved in the process of q-extension of the rational (classical) formulas in enumerative combinatorics, theory of special functions, representation theory, study of integrable systems, and so on. Originating from the paper of Date, Jimbo, Kuniba, Miwa, and Okado on the exactly solvable statistical-mechanics models using the theta function identities [8], the formulas obtained at the q-level are now extended to the elliptic level in many research fields in mathematics and theoretical physics.

In the present monograph, I will show the recent progress of the elliptic extensions in the study of statistical and stochastic models in equilibrium and nonequilibrium statistical mechanics and probability theory. At the elliptic level, many special functions are used, which include Jacobi's theta functions, Weierstrass' elliptic functions, Jacobi's elliptic functions, and others. However, I do not want to make this monograph a handbook of mathematical formulas of these elliptic functions. Hence I have decided to use only the theta function $\theta(\zeta; p)$ of an argument $\zeta \in \mathbb{C}$ and a real nome $p \in (0, 1)$, which is a simplified version of the four kinds of Jacobi's theta functions.

Then we regard θ as a single variable and introduce 'polynomials of θ' with the degree, say $n \in \mathbb{N}$. Instead of considering usual polynomials consisting of $\{\theta^j\}_{j=0}^n$, we consider the function space spanned by n independent functions $\psi_j^{R_n}$, $j = 1, \ldots n$, which can be written using a polynomial of the argument ζ multiplied by a single θ or pairs of such functions (without θ^j with any power $j \geq 2$). The functions $\{\psi_j^{R_n}\}_{j=1}^n$, $n \in \mathbb{N}$ were introduced by Rosengren and Schlosser [64], in association with the

---

[1] Throughout the present monograph, 'elliptic' is used to indicate the complex functions which are meromorphic and doubly periodic. Notice that this word has different meanings in mathematics, for instance, in the field of partial differential equations. There 'elliptic' represents a property related to the two eigenvalues of a certain diagonalizable real matrix: If the two eigenvalues have the same sign, the system is said to be elliptic. In other cases, it could be hyperbolic or parabolic.

seven irreducible reduced affine root systems denoted by $R_n = A_{n-1}, B_n, B_n^\vee, C_n$ $C_n^\vee, BC_n$ and D_n.

Using the R_n theta functions $\{\psi_j^{R_n}\}_{j=1}^n$ of Rosengren and Schlosser and their extensions denoted by $\{\Psi_j^{R_n}\}_{j=1}^n, n \in \mathbb{N}$ in this monograph, we discuss two groups of interacting particle systems, in which n indicates the total number of particles in each system and R specifies one of the seven types for each model. The first group of systems describes the noncolliding Brownian bridges on a one-dimensional torus $\mathbb{T}$ (*i.e.*, a circle) or an interval, and the second one the determinantal point processes (DPPs) on a two-dimensional torus $\mathbb{T}^2$. The former systems are $(1 + 1)$-dimensional stochastic processes and the latter ones are statistical models of stationary point configurations in two dimensions, both of which provide mathematical models for systems of physical particles called fermions, which are interacting with repulsive forces. The boundary conditions and the initial/final configurations of the noncolliding Brownian bridges, and the periodicity conditions of the DPPs are systematically changed depending on the choice of R from the seven types. Moreover, we can argue the scaling limits associated with $n \to \infty$ and precisely define the infinite particle systems. Such limit transitions will be regarded as the mathematical realizations of thermodynamic limits or hydrodynamic limits which are central subjects of statistical mechanics. I would like to emphasize the fact that the construction and analysis of these systems can be performed using the orthogonality properties of $\{\psi_j^{R_n}\}_{j=1}^n$ and $\{\Psi_j^{R_n}\}_{j=1}^n, n \in \mathbb{N}$ with respect to the suitable inner products [38, 39].

I believe that, thanks to the R_n theta functions of Rosengren and Schlosser, here I can show interesting aspects of elliptic extensions in statistical and stochastic systems using only one kind of special function θ.

I would like to thank Christian Krattenthaler very much for his hospitality at Fakultät für Mathematik, Universität Wien, where the present study was started on the sabbatical leave from Chuo University. I express my gratitude to Michael Schlosser, Tomoyuki Shirai, Peter J. Forrester, Piotr Graczyk, Jacek Małecki, Takuya Murayama, and Taiki Endo for their valuable discussion. I also thank Syota Esaki and Shinji Koshida for their careful reading of the draft and for providing very useful comments. All suggestions given by two anonymous reviewers of the manuscript are very important and useful for improving the text and I appreciate their efforts very much. I am grateful to Gernot Akemann, Yacin Ameur, Sungsoo Byum, Nizar Demni, Nam-Gyu Kang, Taro Kimura, Hirofumi Osada, Hjalmer Rosengren, Hideki Tanemura for giving me encouragement to prepare the manuscript. I thank Masayuki Nakamura at the Editorial Department of Springer Japan for his truly kind assistance during the preparation of this monograph. The study reported in this monograph was supported by the Grant-in-Aid for Scientific Research (C) (No.26400405), (No.19K03674), (B) (No.18H01124), (A) (No.21H04432), and (S) (No.16H06338) of the Japan Society for the Promotion of Science (JSPS). This work was supported also by the JSPS Grant-in-Aid for Transformative Research Areas (A) JP22H05105.

Tokyo, Japan   Makoto Katori
February 2023

Contents

Notation

$\mathbf{1}(\omega)$	Indicator of ω; $\mathbf{1}(\omega) = 1$ if ω is satisfied, $\mathbf{1}(\omega) = 0$ otherwise
$\langle \cdot, \cdot \rangle$	$\langle \xi, \phi \rangle := \int_S \phi(x)\xi(dx) = \sum_{j=1}^n \phi(x_j)$ for $\xi \in \mathrm{Conf}(S)$, $\phi \in \mathcal{B}_c(S)$
$\langle \cdot, \cdot \rangle_{\mathbb{T}}$	Inner product for $(\mathcal{E}_{p,r}^{A_{n-1}}, \mathcal{E}_{\hat{p},\hat{r}}^{A_{n-1}})$
$\langle \cdot, \cdot \rangle_{[0,\pi]}$	Inner product for $(\mathcal{E}_p^{R_n}, \mathcal{E}_{\hat{p}}^{R_n})$, $R_n = B_n$, $B_n^\vee$, C_n, $C_n^\vee$, BC_n, D_n
$\langle \cdot, \cdot \rangle_{D(2\pi, 2\pi\mid\tau\mid)}$	Inner product defined by the integral on the fundamental domain $D(2\pi, 2\pi\mid\tau\mid)$ in $\mathbb{C}$
$(\Omega, \mathcal{F}, \mathrm{P})$	Probability space of the one-dimensional Brownian motion (Bm)
$(\Xi, \mathbf{p}, \lambda)$	Point process Ξ specified by the probability density $\mathbf{p}$ with respect to the reference measure λ
(Ξ, K, λ)	Determinantal point process (DPP) Ξ specified by the correlation kernel K with respect to the reference measure λ
$(\Xi_T^{R_n}, \mathbf{K}_T^{R_n}, \lambda^{R_n}(dx))$	Time-dependent DPP of type R_n associated with the noncolliding Brownian bridges (Bbs) with time duration T
$(\Xi^{R_n}, K^{R_n}, \lambda^{R_n}(dx))$	One-dimensional stationary DPP, $R_n = A_{n-1}, B_n, C_n, D_n$
$(\Xi_T^R, \mathcal{K}_T^R, \lambda^R(dx))$	Time-dependent DPP with an infinite number of particles, $R = A, B, C, D$
$(\Xi_{\mathbb{T}^2}^{R_n}, K_{\mathbb{T}^2}^{R_n}, dz)$	DPP on $\mathbb{T}^2$ of type R_n
$(\Xi_{\mathrm{Ginibre}}^R, \mathcal{K}_{\mathrm{Ginibre}}^R, \lambda_{\mathrm{N}})$	Ginibre DPP on $\mathbb{C}$ of type R, $R = A, C, D$

$(\Xi_\alpha, K_\alpha, \lambda_\alpha(dx)) \overset{\alpha \to \infty}{\Rightarrow} (\Xi, K, \lambda(dx))$	Weak convergence of DPP in the vague topology
a.s.	Almost surely
$B(t)$	One-dimensional standard Brownian motion (Bm) at time $t \geq 0$
$\mathcal{B}(\mathbb{R})$	Smallest σ-field containing all intervals on $\mathbb{R}$ (Borel σ-field on $\mathbb{R}$)
$\mathcal{B}_c(S)$	Set of all bounded measurable complex functions with compact support on S
$\mathbb{C}$	Collection of all complex numbers = complex plane, $\mathbb{C} \simeq \mathbb{R}^2$
$\mathbb{C}^\times$	Complex plane punctured at the origin, $\mathbb{C} \backslash \{0\}$
$C_c(S)$	Set of all continuous real-valued functions with compact support on S
$\mathfrak{c}$	Cyclic permutation $\mathfrak{c} = \begin{pmatrix} a\,b\cdots\omega \\ b\,c\cdots a \end{pmatrix}$
$c \circ (\Xi, K, \lambda)$	Dilatation of DPP (Ξ, K, λ) by factor c
$\mathrm{Conf}(S)$	Configuration space of point processes
$\chi(\cdot)$	Test function
$D_{2\pi, 2\pi|\tau|}$	Fundamental domain in $\mathbb{C}$; $D_{2\pi, 2\pi|\tau|} := \{z = x + iy \in \mathbb{C} : 0 \leq x < 2\pi, 0 \leq y < 2\pi|\tau|\}$
δ_{ij}	Kronecker's delta
$\delta(\cdot)$	Dirac's delta function
δ_x	Delta (Dirac) measure with a point mass at x
dx	Lebesgue measure on $\mathbb{R}$
$d\mathbf{x}$	Lebesgue measure on $\mathbb{R}^n$; $d\mathbf{x} := \prod_{j=1}^n dx_j$
dz	Lebesgue measure on $\mathbb{C}$; $dz := d\,\mathrm{Re}\,z\,d\,\mathrm{Im}\,z$
det	Determinant
Det	Fredholm determinant
$\mathbf{E}$	Expectation for the probability measure $\mathbf{P}$
$\mathcal{E}_{p,r}^{A_{n-1}}$	Space of all A_{n-1} theta functions with nome p and norm r
$\mathcal{E}_p^{R_n}$	Space of all R_n theta functions with nome p, $R_n = B_n, B_n^\vee, C_n, C_n^\vee, BC_n, D_n$
$i := \sqrt{-1}$	Imaginary unit
I	$n \times n$ unit matrix
$\mathrm{Im}\,z$	Imaginary part of $z \in \mathbb{C}$
$K(\cdot, \cdot)$	Correlation kernel of DPP

$\mathcal{K}_{\sin}$	Sine (sinc) kernel for the Gaussian unitary ensemble (GUE)
$\mathcal{K}_{\mathrm{chGUE}(\pm 1/2)}$	Correlation kernel for the chiral Gaussian unitary ensemble (chGUE) with $\nu = \pm 1/2$
$\ell(\sigma)$	Number of cyclic permutations in σ
$\overset{\text{(law)}}{=}$	Equivalence in probability law
λ	Reference measure
$\lambda_{\mathbb{T}}$	Uniform measure on $\mathbb{T}$; $\lambda_{\mathbb{T}}(dx) = dx/2\pi$
$\lambda_{[0,1]}$	Uniform measure in an interval $[0, \pi]$; $\lambda_{[0,\pi]}(dx) = dx/\pi$
$\lambda(d\mathbf{x})$	n-direct product measure; $\lambda(d\mathbf{x}) := \prod_{j=1}^{n} \lambda(dx_j)$
$\lambda_{\mathrm{N}}(dz)$	Complex standard normal distribution on $\mathbb{C}$; $\lambda_{\mathrm{N}}(dz) := e^{-|z|^2} dz/\pi$
$M^{R_n}(\zeta; p)$	Macdonald denominator of type R_n, $\zeta \in \mathbb{C}^n$
$\mathbb{N}$	Set of all positive integers, $\{1, 2, \dots\}$
$\mathbb{N}_0$	Set of all nonnegative integers, $\{0, 1, 2, \dots\}$
$(p; p)_\infty$	Special case of p-Pochhammer symbol given by $\prod_{j=1}^{\infty}(1 - p^j)$
$\mathbf{p}$	Probability density for the probability measure $\mathbf{P}$
$p_t, \tilde{p}_t$	$p_t := e^{-t}$, $\tilde{p}_t := e^{-4\pi^2/t}$
$\tilde{p}$	$\tilde{p} := e^{-2\pi i/\tau}$ for $p = e^{2\pi i t}$
$\mathrm{p}(s, x; t, y)$	Transition probability density (tpd) of Bm for $(s, x) \rightsquigarrow (t, y)$
$\mathrm{p}_{\mathbb{T}}$	tpd of Bm on $\mathbb{T}$
p^{aa}	tpd of Bm in $[0, \pi]$ with the absorbing boundary conditions both at 0 and π
p^{ar}	tpd of Bm in $[0, \pi]$ with the absorbing (resp. reflecting) boundary condition at 0 (resp. π)
p^{ra}	tpd of Bm in $[0, \pi]$ with the reflecting (resp. absorbing) boundary condition at 0 (resp. π)
p^{rr}	tpd of Bm in $[0, \pi]$ with the reflecting boundary conditions both at 0 and π
$\tilde{\mathrm{p}}_{\mathbb{T}}$	$\tilde{\mathrm{p}}_{\mathbb{T}}(s, x; t, y) := \sum_{w=-\infty}^{\infty}(-1)^w \mathrm{p}(s, x; t, y + 2\pi w)$
$\mathbf{p}_T^{R_n}$	Probability density of the noncolliding Bbs of type R_n with time duration T

$\{\psi_j^{R_n}\}_{j=1}^n$	R_n theta functions of Rosengren and Schlosser
$\{\Psi_j^{R_n}\}_{j=1}^n$	Orthonormal R_n theta functions having doubly-quasi-periodicity on $\mathbb{C}$
$\Psi[\phi; \kappa]$	Characteristic function of a point process
$\mathbb{R}$	Collection of all real numbers
$\mathbb{R}_+$	Collection of all nonnegative real numbers; $\mathbb{R}_+ := [0, \infty)$
r_t	If n is odd, $r_t = r_t(n) := -p_{n^2 t/2}$; if n is even, $r_t = r_t(n) := p_{n(n+1)t/2}$
$r = (-1)^n p^{n/2}$	Norm for $\{\Psi_j^{A_{n-1}}\}_{j=1}^n$ and $(\Xi_{\mathbb{T}^2}^{A_{n-1}}, K_{\mathbb{T}^2}^{A_{n-1}}, dz)$
Re z	Real part of $z \in \mathbb{C}$
$\mathcal{R}(\Xi, K, \lambda)$	Reflection of DPP (Ξ, K, λ)
ρ_1	First correlation function = density of points
ρ_m	m-point correlation function
S	Subset of $\mathbb{R}^d$
$\mathfrak{S}_n$	Set of all permutations of $\{1, \dots, n\}$ = symmetry group
$S_u(\Xi, K, \lambda)$	Shift of DPP (Ξ, K, λ) by $u \in \mathbb{C}$
$\mathrm{sgn}(\sigma)$	Signature of permutation σ
$\mathrm{SO}(n)$	Special orthogonal group
$\mathrm{Sp}(n)$	Symplectic group
$\mathbb{T}$	One-dimensional torus = unit circle; $\mathbb{T} \simeq \mathbb{R}/2\pi\mathbb{Z}$
$\mathbb{T}^2$	Two-dimensional torus; $\mathbb{T}^2 \simeq (\mathbb{R}/2\pi\mathbb{Z}) \times (\mathbb{R}/2\pi|\tau|\mathbb{Z})$
$\theta(z; p)$	Theta function with argument z and nome p
$\vartheta_j(\xi; \tau)$	Four kinds of Jacobi's theta functions with argument ξ and nome modular parameter τ, $j = 0, 1, 2, 3$
$\mathrm{U}(n)$	Unitary group
$W^{R_n}(\zeta)$	Weyl denominator of type R_n, $\zeta \in \mathbb{C}^n$
$(\mathbf{X}^{R_n}(t))_{t \in [0,T]}$	Noncolliding Brownian bridges (Bbs) of type R_n with n particles and time duration T
$\mathbb{Z}$	Set of all integers $\{\dots, -1, 0, 1, 2, \dots\}$

Chapter 1
Introduction

Abstract The q-extension is the procedure to replace mathematical symbols, identities, functions and others by their meaningful q-analogues, and has been extensively studied. As a typical example of q-extension, here we introduce the q-Pochhammer symbol as the q-analogue of the well-known Pochhammer symbol. The main topic of the present monograph is not the q-extension, but its further extension; the elliptic extension. We introduce a parameter $p \in \mathbb{C}$, $0 < |p| < 1$ instead of q, which is called a nome, and then define the theta function parameterized by p, $\theta(z; p)$, on the complex plane punctured at the origin, $z \in \mathbb{C}^\times$. Basic properties of the theta function used throughout this monograph are summarized here.

1.1 q-Extensions

Let $\mathbb{R}$ be a collection of all real numbers and $\mathbb{C}$ be of all complex numbers, respectively. The imaginary unit is denoted by $i := \sqrt{-1}$. We fix a parameter $q \in \mathbb{C}$ so that $|q| < 1$. For each positive integer $n \in \mathbb{N} := \{1, 2, \dots\}$, the q-analogue $[n]_q$ is defined by

$$[n]_q := \frac{1 - q^n}{1 - q} = 1 + q + \cdots + q^{n-1}.$$

Replacing every occurrence of a positive integer by its q-analogue, the q-analogue of the factorial $n! := n \cdot (n - 1) \cdots 2 \cdot 1$ is given by

$$[n]_q! := [n]_q[n-1]_q \cdots [2]_q[1]_q, \quad n \in \mathbb{N}. \tag{1.1}$$

For convenience, we assume $[0]_q! = 1$. Let $\mathbb{N}_0 := \{0, 1, 2, \dots\}$. Then the q-analogues of the binomial coefficient $\binom{n}{k} = \dfrac{n!}{k!(n - k)!}$ are defined by

$$\begin{bmatrix} n \\ k \end{bmatrix}_q := \frac{[n]_q!}{[k]_q![n - k]_q!}, \quad n \in \mathbb{N}_0, \quad k \in \{0, 1, \dots, n\}. \tag{1.2}$$

The mathematical symbols (1.1) and (1.2) are called the *q-factorial* and the *q-binomial coefficient*, respectively. For $n \in \mathbb{N}_0$, $k \in \{0, 1, \ldots, n\}$, the *q*-analogues of Pascal's triangle relation are given as

$$\begin{bmatrix} n \\ k \end{bmatrix}_q = \begin{bmatrix} n-1 \\ k \end{bmatrix}_q + q^{n-k} \begin{bmatrix} n-1 \\ k-1 \end{bmatrix}_q,$$

$$\begin{bmatrix} n \\ k \end{bmatrix}_q = q^k \begin{bmatrix} n-1 \\ k \end{bmatrix}_q + \begin{bmatrix} n-1 \\ k-1 \end{bmatrix}_q,$$

in which some powers of q should be inserted at the appropriate places. Such procedures to replace mathematical symbols, identities, functions, and so on by their meaningful *q*-analogies are generally called *q-extensions* [6, 21, 51].

The main topic of the present monograph is not the *q*-extension, but its further extension, which is called an *elliptic extension*. In order to pass from *q*-extensions to elliptic extensions smoothly in the next section, here we introduce the *q*-analogues of the Pochhammer symbol $(a)_n := a(a+1) \cdots (a+n-1)$, $n \in \mathbb{N}$, $(a)_0 := 1$ as follows:

$$(a; q)_\infty := \prod_{j=0}^{\infty} (1 - aq^j),$$

$$(a; q)_n := \frac{(a; q)_\infty}{(aq^n; q)_\infty} = \prod_{j=0}^{n-1} (1 - aq^j), \quad n \in \mathbb{N}, \tag{1.3}$$

and $(a; q)_0 := 1$. The q factorials and the *q*-Pochhammer symbols are related as follows:

$$[n]_q! := \frac{(q; q)_n}{(1-q)^n}, \quad n \in \mathbb{N}_0.$$

We notice that the factor $(1 - q)^n$ above is irrelevant when the *q*-binomial coefficient (1.2) is considered:

$$\begin{bmatrix} n \\ k \end{bmatrix}_q = \frac{(q; q)_n}{(q; q)_k (q; q)_{n-k}}, \quad n \in \mathbb{N}_0, \quad k \in \{0, 1, \ldots, n\}.$$

The following abbreviation for a product of the *q*-Pochhammer symbols is used:

$$(a_1, \ldots, a_m; q)_\infty := \prod_{k=1}^{m} (a_k; q)_\infty, \quad m \in \mathbb{N}. \tag{1.4}$$

By the definition (1.3), the following equalities are readily obtained:

$$(a; q)_\infty = \prod_{j=0}^{k-1} (aq^j; q^k)_\infty, \quad (a^k; q^k)_\infty = \prod_{j=0}^{k-1} (a\omega_k^j; q)_\infty, \quad k \in \mathbb{N}, \tag{1.5}$$

where ω_k denotes a primitive k-th root of unity.

1.2 Theta Functions and Elliptic Extensions

We write the complex plane which is punctured at the origin as

$$\mathbb{C}^\times := \mathbb{C} \setminus \{0\} = \{z \in \mathbb{C} : 0 < |z| < \infty\}.$$

Let $p \in \mathbb{C}$ be a fixed number so that $0 < |p| < 1$. The *theta function* with argument z and *nome* p is defined by[1]

$$\theta(z; p) := (z, p/z; p)_\infty$$

$$= \prod_{j=0}^{\infty} (1 - zp^j)(1 - p^{j+1}/z). \tag{1.6}$$

We often use the shorthand notation $\theta(z_1, \ldots, z_m; p) := \prod_{k=1}^{m} \theta(z_k; p), m \in \mathbb{N}$.

By this definition, we can readily see that

$$\lim_{p \to 0} \theta(z; p) = 1 - z. \tag{1.7}$$

It implies that

$$\lim_{p \to 0} \frac{\theta(q^n; p)}{1 - q} = [n]_q, \quad n \in \mathbb{N},$$

that is, the $p \to 0$ limit of $\theta(q^n; p)/(1 - q)$ is the q-analogue $[n]_q$ of a positive integer $n \in \mathbb{N}$. Conversely, we may say that $\theta(q^n; p)/(1 - q)$ can be regarded as an extension of $[n]_q$ by introducing another parameter p in addition to q. Moreover, if we consider $\theta(aq^n; p)$ with $a = e^{-2i\alpha}$ and $q = e^{-2i\phi}$, $\alpha, \phi \in [0, 2\pi)$, then (1.7) proves

$$\lim_{p \to 0} \frac{\theta(aq^n; p)}{2i\sqrt{aq^n}} = \sin(\alpha + n\phi).$$

---

[1] This function is sometimes called the *modified theta function*. The relations with the four types of Jacobi's theta functions are given in Sect. 2.4. In this monograph, we will simply call $\theta(z; p)$ the *theta function*.

This suggests that the present extension expressed by theta functions should be regarded as being at the higher level than the trigonometric functions (or the hyperbolic functions). For this reason, this second extension following the q-extension will be called the *elliptic extension*.[2]

As a function of z, the theta function $\theta(z; p)$ is holomorphic in $\mathbb{C}^\times$ and has only single zeros precisely at p^j, $j \in \mathbb{Z} := \{0, \pm 1, \pm 2, \ldots\}$, that is, the zero set is given by

$$\{z \in \mathbb{C}^\times : \theta(z; p) = 0\} = \{p^j : j \in \mathbb{Z}\}. \tag{1.8}$$

By the definition (1.6) the following equalities are proved:

$$\theta(1/z; p) = -\frac{1}{z}\theta(z; p), \tag{1.9}$$

$$\theta(pz; p) = -\frac{1}{z}\theta(z; p), \tag{1.10}$$

which are referred to as the *inversion formula* and the *quasi-periodicity*, respectively. (See Exercise 1.1.) By comparing (1.9) and (1.10) using the transformation $z \mapsto 1/z$, we immediately see the *symmetry*,

$$\theta(p/z; p) = \theta(z; p). \tag{1.11}$$

Corresponding to (1.5), the following equalities hold:

$$\theta(z; p) = \prod_{j=0}^{k-1} \theta(zp^j; p^k), \quad \theta(z^k; p^k) = \prod_{j=0}^{k-1} \theta(z\omega_k^j; p), \quad k \in \mathbb{N}, \tag{1.12}$$

where ω_k denotes a primitive k-th root of unity. (See Exercise 1.2.)

In $\mathbb{C}^\times$, $\theta(z; p)$ is holomorphic, and it allows a Laurent expansion,

$$\theta(z; p) = \frac{1}{(p; p)_\infty} \sum_{n \in \mathbb{Z}} (-1)^n p^{\binom{n}{2}} z^n, \tag{1.13}$$

where $\binom{n}{2} := n(n-1)/2, n \in \mathbb{Z}$. Since the theta function is defined as the product of two q-Pochhammer symbols as (1.6), the above is written as

---

[2] Doubly periodic functions which are analytic in any finite domain on $\mathbb{C}$ except for poles (i.e., meromorphic) are generally called *elliptic functions* [57, 78]. Consider $\theta(e^{i\varphi}; p)$ by putting $p = e^{2\pi i \tau}$ with Im $\tau > 0$. The periodicity in $\varphi \mapsto \varphi + 2\pi$ is obvious. By (1.11), this has another invariance under $\varphi \mapsto -\varphi + 2\pi\tau$. The theta function is not elliptic, but as suggested by the above observations, elliptic functions can be constructed as rational functions of θ's. The notion of elliptic extension is much wider and deeper than what we will explain in this monograph, but we can say that the theta function plays important roles everywhere. For advanced studies of elliptic extensions, see, for instance, [2, 4, 7–9, 21, 27, 29, 30, 44, 48, 50, 62, 64, 67, 72–74, 77] and references therein.

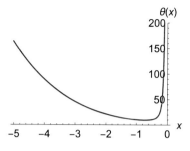
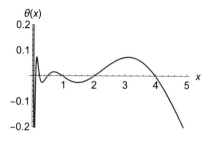

Fig. 1.1 For $p = 1/2$, the theta function $\theta(x; p)$ is plotted for the negative argument $x < 0$ in the left figure and for the positive argument $x > 0$ in the right figure, respectively. When $x < 0$, the theta function is positive definite, but when $x > 0$ it oscillates around zero infinitely many times as shown by (1.17)

$$\sum_{n \in \mathbb{Z}} (-1)^n p^{\binom{n}{2}} z^n = (p; p)_\infty (z; p)_\infty (p/z; p)_\infty,$$

which is known as *Jacobi's triple product identity*.

One can show that

$$\theta'(1; p) := \left. \frac{\partial \theta(z; p)}{\partial z} \right|_{z=1} = -(p; p)_\infty^2. \tag{1.14}$$

The theta function satisfies the *Riemann–Weierstrass addition formula* [49],

$$\theta(xy, x/y, uv, u/v; p) - \theta(xv, x/v, uy, u/y; p) = \frac{u}{y} \theta(yv, y/v, xu, x/u; p),$$
$$\tag{1.15}$$

which is the elliptic extension of the addition formula of trigonometric functions. (See Exercise 1.3.)

When $p \in (0, 1)$, we see that

$$\overline{\theta(z; p)} = \theta(\overline{z}; p). \tag{1.16}$$

In this case the definition (1.6) implies the following (see Fig. 1.1):

$$\left. \begin{array}{l} \theta(x; p) > 0, \ x \in (p^{2j+1}, p^{2j}) \\ \theta(x; p) = 0, \ x = p^j \\ \theta(x; p) < 0, \ x \in (p^{2j}, p^{2j-1}) \end{array} \right\} \ j \in \mathbb{Z},$$
$$\theta(x; p) > 0, \quad x \in (-\infty, 0). \tag{1.17}$$

Moreover, we can prove the following. In the interval $x \in (-\infty, 0)$, $\theta(x; p)$ is strictly convex with

$$\min_{x \in (-\infty, 0)} \theta(x; p) = \theta(-\sqrt{p}; p) = \prod_{n=1}^{\infty} (1 + p^{n-1/2})^2 > 0, \tag{1.18}$$

Fig. 1.2 For $p = 1/4$, the theta function $\theta(x; p)$ is plotted for $x < 0$. It is strictly convex and at $x = -\sqrt{1/4} = -0.5$ it attains its minimum which is positive

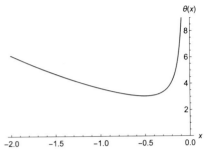

Fig. 1.3 For $p = 1/4$, the theta function $\theta(x; p)$ is plotted for $x \in (p, 1) = (0.25, 1)$. It is strictly concave and at $x = \sqrt{1/4} = 0.5$ it attains its maximum. At both edges of the interval, $x = p = 0.25$ and $x = 1$, $\theta(x)$ becomes zero

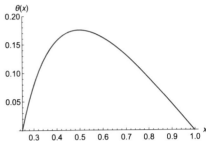

and $\lim_{x \downarrow -\infty} \theta(x; p) = \lim_{x \uparrow 0} \theta(x; p) = +\infty$, and in the interval $x \in (p, 1), \theta(x; p)$ is strictly concave with

$$\max_{x \in (p,1)} \theta(x; p) = \theta(\sqrt{p}; p) = \prod_{n=1}^{\infty}(1 - p^{n-1/2})^2, \tag{1.19}$$

$\theta(x; p) \sim (p; p)_\infty^2 (x - p)/p$ as $x \downarrow p$, and $\theta(x; p) \sim (p; p)_\infty^2 (1 - x)$ as $x \uparrow 1$, where (1.10) and (1.14) were used. See Figs. 1.2 and 1.3.

Exercises

1.1 By the definition of the theta function (1.6), prove the inversion formula (1.9) and the quasi-periodicity (1.10).

1.2 By direct calculation, verify the following equalities:

$$\theta(\zeta; p^3)\theta(\zeta p; p^3)\theta(\zeta p^2; p^3) = \theta(\zeta; p), \tag{1.20}$$

$$\theta(\zeta; p)\theta(\zeta \omega_3; p)\theta(\zeta \omega_3^2; p) = \theta(\zeta^3; p^3), \tag{1.21}$$

where ω_3 is a primitive 3rd root of unity.

1.3 Let $x = e^{-2ia}$, $y = e^{-2ib}$, $u = e^{-2ic}$, and $v = e^{-2id}$ in the Riemann–Weierstrass addition formula (1.15). Consider the limit $p \to 0$ and derive the following trigonometric equality:

$$
\begin{aligned}
& \sin(a+b)\sin(a-b)\sin(c+d)\sin(c-d) \\
& - \sin(a+d)\sin(a-d)\sin(c+b)\sin(c-b) \\
& = \sin(b+d)\sin(b-d)\sin(a+c)\sin(a-c).
\end{aligned}
\tag{1.22}
$$

Show that (1.22) is readily verified if we use the following addition formulas of trigonometric functions:

$$
\sin(a \pm b) = \sin a \cos b \pm \cos a \sin b.
\tag{1.23}
$$

Chapter 2
Brownian Motion and Theta Functions

Abstract We introduce the Brownian motion on a real line $\mathbb{R}$. First we notice that its transition probability density solves the heat equation starting from a single delta function. Then we consider the Brownian motion on a unit circle, which is regarded as a one-dimensional torus and is denoted by $\mathbb{T}$. Two different formulas of the transition probability are given, both of which are expressed using the theta function with different nomes. The equivalence of these two expressions implies Jacobi's imaginary transformation of the theta function. We also study the Brownian motion on a semi-circle which is identified with the interval $[0, \pi]$ with two boundary points 0 and π. We impose the absorbing boundary condition or the reflecting boundary condition at each of the boundary points and hence we obtain four types of Brownian motion in the interval. We see an interesting correspondence between these four types of Brownian motion and the four types of Jacobi's theta functions via expressions of the transition probability densities.

2.1 Brownian Motion on $\mathbb{R}$

The one-dimensional real space is denoted by $\mathbb{R}$. We consider the motion of a Brownian particle in $\mathbb{R}$ starting from the origin 0 at time $t = 0$. Each realization of its random trajectory is called a *sample path*, denoted by ω. Let Ω be the collection of all sample paths and call it the *sample path space*. In a path $\omega \in \Omega$ the position of the Brownian particle at each time $t \geq 0$ is written as $B(t, \omega)$. Usually we omit ω and simply write it as $B(t)$, $t \geq 0$.

Each event associated with the process is represented by a subset of Ω, and the collection of all events is denoted by $\mathcal{F}$. The whole sample path space Ω and the empty set $\emptyset$ are in $\mathcal{F}$. For any two sets $\mathsf{A}, \mathsf{B} \in \mathcal{F}$, we assume that $\mathsf{A} \cup \mathsf{B} \in \mathcal{F}$ and $\mathsf{A} \cap \mathsf{B} \in \mathcal{F}$. If $\mathsf{A} \in \mathcal{F}$, then its complement $\mathsf{A}^{c} := \Omega \setminus \mathsf{A}$ is also in $\mathcal{F}$. It is closed for any infinite sum of events in the sense that, if $\mathsf{A}_n \in \mathcal{F}$, $n = 1, 2, \ldots$, then $\cup_{n \geq 1} \mathsf{A}_n \in \mathcal{F}$. Such a collection is said to be a σ-*field*.

A *probability measure* P is a function defined on the σ-field $\mathcal{F}$. Since any element of $\mathcal{F}$ is given by a set as above, any input of P is a set, and hence P is a set function. It satisfies the following properties; $\mathsf{P}[\mathsf{A}] \geq 0$ for all $\mathsf{A} \in \mathcal{F}$,

© The Author(s), under exclusive license to Springer Nature Singapore Pte Ltd. 2023 9
M. Katori, *Elliptic Extensions in Statistical and Stochastic Systems*,
SpringerBriefs in Mathematical Physics 47,
https://doi.org/10.1007/978-981-19-9527-9_2

$P[\Omega] = 1$, $P[\emptyset] = 0$, and if $A_n \in \mathcal{F}$, $n \in \mathbb{N}$, are disjoint; $A_n \cap A_m = \emptyset$, $n \neq m$, then $P[\bigcup_{n=1}^{\infty} A_n] = \sum_{n=1}^{\infty} P[A_n]$. In particular, $P[A^c] = 1 - P[A]$ for all $A \in \mathcal{F}$. The triplet $(\Omega, \mathcal{F}, P)$ is called the *probability space*.

The smallest σ-field containing all intervals on $\mathbb{R}$ is called the Borel σ-field and denoted by $\mathcal{B}(\mathbb{R})$. A *random variable* is a measurable function $f(\omega)$ on Ω such that, for every Borel set $A \in \mathcal{B}(\mathbb{R})$, $f^{-1}(A) \in \mathcal{F}$. Two events A and B are said to be *independent* if $P[A \cap B] = P[A]P[B]$. Two random variables X and Y are *independent* if the events $A = \{X : X \in A\}$ and $B = \{Y : Y \in B\}$ are independent for any $A, B \in \mathcal{B}(\mathbb{R})$.

The *one-dimensional standard Brownian motion*, $\{B(t, \omega) : t \geq 0\}$, has the following three properties:

(BM1) $P[B(0, \omega) = 0] = 1$.

(BM2) There is a subset $\widetilde{\Omega} \subset \Omega$, such that $P[\widetilde{\Omega}] = 1$ and for any $\omega \in \widetilde{\Omega}$, $B(t, \omega)$ is a real continuous function of t. We say that $B(t)$ has a *continuous path* almost surely (a.s., for short).

(BM3) For an arbitrary $n \in \mathbb{N} := \{1, 2, 3, \dots \}$, and for any sequence of times, $t_0 := 0 < t_1 < \cdots < t_n$, the increments $B(t_m) - B(t_{m-1})$, $m = 1, 2, \dots, n$, are independent, and each increment is in the *normal distribution* (*Gaussian distribution*) with mean 0 and variance $\sigma^2 = t_m - t_{m-1}$. It means that for any $1 \leq m \leq M$ and $a < b$,

$$P[B(t_m) - B(t_{m-1}) \in [a, b]] = \int_a^b \mathrm{p}(t_{m-1}, 0; t_m, x) dx,$$

where for $0 \leq s \leq t < \infty$, $x, y \in \mathbb{R}$ we define

$$\mathrm{p}(s, x; t, y) := \begin{cases} \dfrac{1}{\sqrt{2\pi(t-s)}} e^{-(y-x)^2/2(t-s)}, & \text{for } t > s, \\ \delta(x-y), & \text{for } t = s. \end{cases} \tag{2.1}$$

It is obvious from (2.1) in the property **(BM3)** that

$$\mathrm{p}(s, x; t, y) = \mathrm{p}(0, 0; t-s, y-x),$$
$$\mathrm{p}(s, x; t, y) = \mathrm{p}(s, y; t, x), \quad 0 \leq s \leq t, \ x, y \in \mathbb{R}. \tag{2.2}$$

By **(BM3)** we see that, for any $0 \leq s < t < \infty$, $x \in \mathbb{R}$, and $A \in \mathcal{B}(\mathbb{R})$,

$$P[B(t) \in A | B(s) = x] = \int_A \mathrm{p}(s, x; t, y) dy,$$

where the LHS denotes the probability for $B(t) \in A$ with the condition that the position of the one-dimensional standard Brownian motion was x at time s. The function $\mathrm{p}(s, x; t, y)$ is then called the *transition probability density* of the one-dimensional standard Brownian motion.[1] It gives the fundamental solution of the *heat equation* (*diffusion equation*) in the sense that

$$\frac{\partial}{\partial t}p(0, x; t, y) = \frac{1}{2}\frac{\partial^2}{\partial y^2}p(0, x; t, y)$$
$$= \frac{1}{2}\frac{\partial^2}{\partial x^2}p(0, x; t, y), \tag{2.3}$$

with

$$\lim_{t\downarrow 0} p(0, x; t, y) = \delta(x - y). \tag{2.4}$$

Moreover, it is easy to verify that the *Chapman–Kolmogorov equation* holds:

$$\int_{\mathbb{R}} p(s, x; t, y)p(t, y; u, z)dy = p(s, x; u, z), \quad 0 \le s \le t \le u, \quad x, z \in \mathbb{R}, \tag{2.5}$$

which implies the *Markov property* of Brownian motion.

For any $c > 0$, the following equality is satisfied by the transition probability (2.1) of the one-dimensional standard Brownian motion:

$$p(c^2 s, cx; c^2 t, cy)d(cy) = p(s, x; t, y)dy. \tag{2.6}$$

This implies the *equivalence in probability law*

$$(B(t))_{t\ge 0} \overset{\text{(law)}}{=} \left(\frac{1}{c}B(c^2 t)\right)_{t\ge 0}, \quad \forall c > 0, \tag{2.7}$$

which is called *Brownian motion scaling*. It is obvious that if we perform a time change $t \to 2\alpha t$, $\alpha > 0$, the obtained transition probability density

$$p(0, x; 2\alpha t, y) = \frac{1}{\sqrt{4\pi \alpha t}}e^{-(y-x)^2/4\alpha t}, \quad t > 0$$

satisfies the heat equation with *diffusivity constant* α,

$$\frac{\partial}{\partial t}p(0, x; 2\alpha t, y) = \alpha\frac{\partial^2}{\partial y^2}p(0, x; 2\alpha t, y).$$

The standard Brownian motion is a special case with $\alpha = 1/2$.

---

[1] By the shift invariance in space and time given by the first equality in (2.2), the property (**BM1**) is readily generalized: The Brownian motion started at an arbitrary fixed point $x \in \mathbb{R}$ a.s. is defined by $B^x(t) := x + B(t)$, $t \ge 0$. The transition probability density for $B^x(t)$ is given by $p(0, x; t, y)$, $t \ge 0$, $y \in \mathbb{R}$.

From now on, we refer to the one-dimensional standard Brownian motion simply as *Brownian motion*.

2.2 Brownian Motion on the One-Dimensional Torus $\mathbb{T}$

We write the unit circle on the complex plane as

$$\mathbb{T} := \{z \in \mathbb{C} : |z| = 1\}.$$

It is a one-dimensional torus; $\mathbb{T} = \mathbb{R}/2\pi\mathbb{Z}$. Each point in $\mathbb{T}$ is expressed by e^{ix}, $x \in [0, 2\pi)$. In the following we will introduce a Brownian motion on $\mathbb{T}$. The transition probability density is denoted by $p_{\mathbb{T}}(s, x; t, y)$ for $0 < s < t < \infty$ and $x, y \in [0, 2\pi)$. This should solve the heat equation (2.3) with (2.4), and satisfy the periodic condition

$$p_{\mathbb{T}}(s, x; t, y + 2\pi n) = p_{\mathbb{T}}(s, x; t, y), \quad \forall n \in \mathbb{Z}.$$

For this periodicity, we assume that it is expressed in the form

$$p_{\mathbb{T}}(0, x; t, y) = \sum_{n=-\infty}^{\infty} C_n(t) \cos(n(y - x))$$

with a series of time-dependent coefficients $(C_n(t))_{n \in \mathbb{Z}}, t \geq 0$. In order to solve (2.3), the time dependence of $C_n(t)$ is determined as $C_n(t) = c_n e^{-n^2 t/2}$. Then the initial condition (2.4) determines $c_n \equiv 1/2\pi$ for all $n \in \mathbb{Z}$, since $\delta(x)$ is expressed as

$$\delta(x) = \frac{1}{2\pi} \sum_{n=-\infty}^{\infty} e^{inx} = \frac{1}{2\pi} \sum_{n=-\infty}^{\infty} \cos(nx), \quad x \in \mathbb{R}.$$

On the other hand, $p_{\mathbb{T}}$ will be obtained by 'wrapping' the transition probability density on $\mathbb{R}$ given by (2.1),

$$p_{\mathbb{T}}(0, x; t, y) = \sum_{w=-\infty}^{\infty} p(0, x; t, y + 2\pi w),$$

where w denotes the *winding number* of the Brownian path around the circle. Hence we have two different expressions for $p_{\mathbb{T}}$:

$$\mathrm{p}_{\mathbb{T}}(0, x; t, y) = \frac{1}{2\pi} \sum_{n=-\infty}^{\infty} e^{-n^2 t/2} \cos(n(y-x))$$

$$= \sum_{w=-\infty}^{\infty} \frac{1}{\sqrt{2\pi t}} e^{-(y-x+2\pi w)^2/2t}, \quad t > 0, \quad x, y \in [0, 2\pi).$$

Compare with the Laurent expansion of the theta function (1.13); the above two expressions are written using the theta functions with different arguments and nomes:

$$\mathrm{p}_{\mathbb{T}}(0, x; t, y) = \frac{(p_t; p_t)_\infty}{2\pi} \theta(-p_t^{1/2} e^{i(y-x)}; p_t)$$

$$= \frac{e^{-(y-x)^2/2t}}{\sqrt{2\pi t}} (\widetilde{p}_t; \widetilde{p}_t)_\infty \theta(-\widetilde{p}_t^{1/2} e^{-2\pi(y-x)/t}; \widetilde{p}_t), \quad x, y \in [0, 2\pi),$$

$$\tag{2.8}$$

where

$$p_t := e^{-t}, \quad \widetilde{p}_t := e^{-4\pi^2/t}, \quad t > 0. \tag{2.9}$$

The above provides a probability theoretical proof of the equality between the theta functions [1], which is known as *Jacobi's imaginary transformation*. The argument $-p_t^{1/2} e^{i(y-x)}$ in the former expression is transformed to $-\widetilde{p}_t^{1/2} e^{-2\pi(y-x)/t}$ in the latter one in (2.8), and hence the coordinates x and y in $\mathbb{T}$ seem to be changed to pure imaginaries $2\pi i x/t$ and $2\pi i y/t$, respectively. Here we give a general form of Jacobi's imaginary transformation as a lemma. For the nome $p \in \mathbb{C}$, $|p| < 1$, we define the *nome modular parameter* τ by

$$p = e^{2\pi i \tau} =: p(\tau). \tag{2.10}$$

By definition, $\tau \in \mathbb{C}$ and $\mathrm{Im}\, \tau > 0$.

Lemma 2.1 *With (2.10), we define*

$$\widetilde{p} := p(-1/\tau) = e^{-2\pi i/\tau}. \tag{2.11}$$

Then, for $\zeta \in \mathbb{C}$, the following equality holds:

$$\theta(e^{i\zeta}; p) = \frac{e^{3\pi i/4}}{\tau^{1/2}} \frac{\widetilde{p}^{1/8}(\widetilde{p}; \widetilde{p})_\infty}{p^{1/8}(p; p)_\infty} \exp\left[-i\frac{\zeta^2}{4\pi\tau} + i\left(1 - \frac{1}{\tau}\right)\frac{\zeta}{2}\right] \theta(e^{-i\zeta/\tau}; \widetilde{p}).$$

$$\tag{2.12}$$

By the above construction, the symmetry (2.2) of the transition probability p of the Brownian motion on $\mathbb{R}$ is inherited as

$$\mathrm{p}_\mathbb{T}(s, x; t, y) = \mathrm{p}_\mathbb{T}(0, 0; t - s, y - x),$$

$$\mathrm{p}_\mathbb{T}(s, x; t, y) = \mathrm{p}_\mathbb{T}(s, y; t, x), \quad \text{for any } 0 \le s \le t, \ x, y \in [0, 2\pi). \quad (2.13)$$

The second equality in the above can be directly proved by applying the symmetry of the theta function (1.11) to the first expression in (2.8) as $\theta(-p_t^{1/2} e^{i(y-x)}; p_t) = \theta(p_t/(-p_t^{1/2} e^{i(y-x)}); p_t) = \theta(-p_t^{1/2} e^{i(x-y)}; p_t)$.

We see a reciprocity of time in (2.9). By the asymptotic (1.7) of the theta function, we see that the former expression is useful to estimate the long-term behavior of the transition probability as

$$\mathrm{p}_\mathbb{T}(0, x; t, y) \sim \frac{1}{2\pi}(1 + p_t^{1/2} e^{i(y-x)}) \sim \frac{1}{2\pi} \quad \text{as } t \to \infty,$$

since $p_t \to 0$ as $t \to \infty$, while the latter is useful for the short-term behavior as

$$\mathrm{p}_\mathbb{T}(0, x; t, y) \sim \frac{e^{-(y-x)^2/2t}}{\sqrt{2\pi t}}(1 + \widetilde{p}_t^{1/2} e^{-2\pi(y-x)/t}) \sim \delta(y - x) \quad \text{as } t \to 0,$$

since $\widetilde{p}_t \to 0$ as $t \to 0$. The above shows that the distribution of the location of $B(t)$ started from a delta measure at x is relaxed to the uniform measure on $\mathbb{T}$ as $t \to \infty$.

The Chapman–Kolmogorov equation (2.5) for p is expressed for $\mathrm{p}_\mathbb{T}$ as

$$\int_\mathbb{T} \mathrm{p}_\mathbb{T}(s, x; t, y)\mathrm{p}_\mathbb{T}(t, y; u, z)dy = \mathrm{p}_\mathbb{T}(s, x; u, z), \quad 0 \le s < t < u, \ x, z \in [0, 2\pi),$$
$$(2.14)$$

and it proves the *convolution formula for the theta function*,

$$\frac{1}{2\pi} \int_0^{2\pi} \theta(-p_{t-s}^{1/2} e^{i(y-x)}; p_{t-s})\theta(-p_{u-t}^{1/2} e^{i(z-y)}; p_{u-t})dy$$

$$= \frac{(p_{u-s}; p_{u-s})_\infty}{(p_{t-s}; p_{t-s})_\infty (p_{u-t}; p_{u-t})_\infty}\theta(-p_{u-s}^{1/2} e^{i(z-x)}; p_{u-s}), \quad (2.15)$$

$0 \le s < t < u, x, z \in [0, 2\pi)$. Here notice the multiplicities in arguments and nomes such that $p_{t-s}^{1/2} e^{i(y-x)} p_{u-t}^{1/2} e^{i(z-y)} = p_{u-s}^{1/2} e^{i(z-x)}$ and $p_{t-s} p_{u-t} = p_{u-s}$, which are guaranteed by our choice (2.9) and are needed to establish (2.15).

2.3 Brownian Motion in the Interval $[0, \pi]$

Now we consider the upper-half circle, $\mathbb{T} \cap \{z \in \mathbb{C} : \mathrm{Im}\, z \ge 0\}$ and identify it with the interval $[0, \pi]$ in $\mathbb{R}$. We consider a Brownian motion in $[0, \pi]$ imposing one of the two kinds of boundary conditions. One of them is the *absorbing boundary condition* for Brownian motion such that, at the boundary point $b \in \{0, \pi\}$, the Brownian motion is killed. The corresponding transition probability density should satisfy the *Dirichlet boundary condition*. Another one is the *reflecting boundary condition* such that, at $b \in \{0, \pi\}$, the transition probability density should satisfy the *Neumann boundary condition*.

2.3.1 Absorbing at Both Boundary Points

The transition probability density denoted by $p^{aa}_{[0,\pi]}(0, x; t, y)$ should solve the heat equation

$$\frac{\partial}{\partial t} u(t, y) = \frac{1}{2}\frac{\partial^2}{\partial y^2} u(t, y), \quad t > 0 \tag{2.16}$$

with the initial condition

$$u(0, y) = \delta(x - y), \quad x, y \in [0, \pi], \tag{2.17}$$

under the Dirichlet boundary condition

$$u(t, 0) = u(t, \pi) = 0, \quad t \geq 0. \tag{2.18}$$

For (2.18), we choose the basis of series expansion as $\{\sin(ny)\}_{n \in \mathbb{Z}}$ and obtain the result,

$$p^{aa}_{[0,\pi]}(0, x; t, y) = \frac{1}{\pi}\sum_{n=-\infty}^{\infty} e^{-n^2 t/2} \sin(nx)\sin(ny), \quad t \geq 0, \ x, y \in [0, \pi]. \tag{2.19}$$

It is easy to confirm that this is written as follows:

$$p^{aa}_{[0,\pi]}(0, x; t, y) = \sum_{w=-\infty}^{\infty} \{p(0, x; t, y + 2\pi w) - p(0, -x; t, y + 2\pi w)\}$$

$$= p_{\mathbb{T}}(0, x; t, y) - p_{\mathbb{T}}(0, -x; t, y)$$

$$= \frac{(p_t; p_t)_\infty}{2\pi}\{\theta(-p_t^{1/2}e^{i(y-x)}; p_t) - \theta(-p_t^{1/2}e^{i(y+x)}; p_t)\}, \tag{2.20}$$

$t > 0, x, y \in [0, \pi]$. The expression (2.19) is called the *spectral representation* of the transition probability density in Appendix 1.6 of [3]. The first expression in (2.20) coincides with the formula given there which is obtained by applying the *reflection principle of Brownian motion*.

2.3.2 Reflecting at Both Boundary Points

The transition probability density denoted by $p_{[0,\pi]}^{\mathrm{rr}}(0, x; t, y)$ should solve (2.16) with the initial condition (2.17) under the Neumann boundary condition

$$\left.\frac{\partial u(t, y)}{\partial y}\right|_{y=0} = \left.\frac{\partial u(t, y)}{\partial y}\right|_{y=\pi} = 0, \quad t \geq 0. \tag{2.21}$$

For (2.21), we choose the basis of series expansion as $\{\cos(ny)\}_{n \in \mathbb{Z}}$ and obtain the result,

$$p_{[0,\pi]}^{\mathrm{rr}}(0, x; t, y) = \frac{1}{\pi} \sum_{n=-\infty}^{\infty} e^{-n^2 t/2} \cos(nx) \cos(ny), \quad t \geq 0, \quad x, y \in [0, \pi]. \tag{2.22}$$

It is easy to confirm that this is written as follows:

$$\begin{aligned} p_{[0,\pi]}^{\mathrm{rr}}(0, x; t, y) &= \sum_{w=-\infty}^{\infty} \{p(0, x; t, y + 2\pi w) + p(0, -x; t, y + 2\pi w)\} \\ &= p_{\mathbb{T}}(0, x; t, y) + p_{\mathbb{T}}(0, -x; t, y) \\ &= \frac{(p_t; p_t)_\infty}{2\pi} \{\theta(-p_t^{1/2} e^{i(y-x)}; p_t) + \theta(-p_t^{1/2} e^{i(y+x)}; p_t)\}, \end{aligned} \tag{2.23}$$

$t > 0, x, y \in [0, \pi]$. The former expression coincide with the formula given in Appendix 1.5 in [3], where the expression (2.22) is the *spectral representation* of the transition probability density.

2.3.3 Absorbing at One Boundary Point and Reflecting at Another Boundary Point

The transition probability density denoted by $p_{[0,\pi]}^{\mathrm{ar}}(0, x; t, y)$ should solve (2.16) with the initial condition (2.17) under the following boundary condition:

$$u(t, 0) = 0, \quad \left.\frac{\partial u(t, y)}{\partial y}\right|_{y=\pi} = 0, \quad t \geq 0. \tag{2.24}$$

For (2.24), we choose the basis of series expansion as $\{\sin\{(n - 1/2)y\}\}_{n \in \mathbb{Z}}$ and obtain the result,

$$p_{[0,\pi]}^{\mathrm{ar}}(0, x; t, y) = \frac{1}{\pi} \sum_{n=-\infty}^{\infty} e^{-(n-1/2)^2 t/2} \sin\{(n - 1/2)x\} \sin\{(n - 1/2)y\}, \tag{2.25}$$

$t \geq 0$, x, $y \in [0, \pi]$. This is written as follows [38]:

$$p_{[0,\pi]}^{\text{ar}}(0, x; t, y) = \frac{p_t^{1/8}(p_t; p_t)_\infty}{2\pi}$$
$$\times \{e^{-i(y-x)/2}\theta(-e^{i(y-x)}; p_t) - e^{-i(y+x)/2}\theta(-e^{i(y+x)}; p_t)\}, \qquad (2.26)$$

$t > 0$, x, $y \in [0, \pi]$.

By changing the variables as $x \to \pi - x$ and $y \to \pi - x$, we can exchange the boundary conditions at 0 and π, and hence $p_{[0,\pi]}^{\text{ra}}(0, x; t, y) = p_{[0,\pi]}^{\text{ar}}(0, \pi - x; t, \pi - y)$. We have the following results:

$$p_{[0,\pi]}^{\text{ra}}(0, x; t, y) = \frac{1}{\pi} \sum_{n=-\infty}^{\infty} e^{-(n-1/2)^2 t/2} \cos\{(n - 1/2)x\} \cos\{(n - 1/2)y\}$$

$$= \frac{p_t^{1/8}(p_t; p_t)_\infty}{2\pi}$$
$$\times \{e^{-i(y-x)/2}\theta(-e^{i(y-x)}; p_t) + e^{-i(y+x)/2}\theta(-e^{i(y+x)}; p_t)\}, \qquad (2.27)$$

$t > 0$, x, $y \in [0, \pi]$.

We introduce the following sum of the transition probability densities on $\mathbb{R}$ with alternating signs:

$$\widetilde{p}_{\mathbb{T}}(0, x; t, y) := \sum_{w=-\infty}^{\infty} (-1)^w p(0, x; t, y + 2\pi w). \qquad (2.28)$$

Using Jacobi's imaginary transformation (2.9), we can rewrite the above as

$$\widetilde{p}_{\mathbb{T}}(0, x; t, y) = \frac{p_t^{1/8}(p_t; p_t)_\infty}{2\pi} e^{-i(y-x)/2}\theta(-e^{i(y-x)}; p_t). \qquad (2.29)$$

Hence (2.26) and (2.27) are expressed as follows:

$$p_{[0,\pi]}^{\text{ar}}(0, x; t, y) = \widetilde{p}_{\mathbb{T}}(0, x; t, y) - \widetilde{p}_{\mathbb{T}}(0, -x; t, y),$$
$$p_{[0,\pi]}^{\text{ra}}(0, x; t, y) = \widetilde{p}_{\mathbb{T}}(0, x; t, y) + \widetilde{p}_{\mathbb{T}}(0, -x; t, y), \quad t > 0, \quad x, y \in [0, \pi]. \qquad (2.30)$$

We notice that $\widetilde{p}_{\mathbb{T}}$ satisfies the Chapman–Kolmogorov equation,

$$\int_{\mathbb{T}} \widetilde{p}_{\mathbb{T}}(s, x; t, y)\widetilde{p}_{\mathbb{T}}(t, y; u, z)dy = \widetilde{p}_{\mathbb{T}}(s, x; u, z), \quad 0 \leq s < t < u, \ x, z \in [0, 2\pi), \qquad (2.31)$$

and $p_{[0,\pi]}^{\sharp}$'s with $\sharp \in \{\text{aa, ar, ra, rr}\}$ also do. It is easy to verify the following symmetry:

$$p^{\sharp}_{[0,\pi]}(s, x; t, y) = p^{\sharp}_{[0,\pi]}(0, 0; t - s, y - x),$$

$$p^{\sharp}_{[0,\pi]}(s, x; t, y) = p^{\sharp}_{[0,\pi]}(s, y; t, x), \quad \text{for any } 0 < s < t, \ x, y \in [0, \pi]. \quad (2.32)$$

The former is obvious by the above definition. The latter is directly shown using the symmetry (1.11) of the theta function to the expressions (2.20) and (2.23) for $p^{aa}_{[0,\pi]}$ and $p^{rr}_{[0,\pi]}$, and the inversion formula (1.9) to (2.26) and (2.27) for $p^{ar}_{[0,\pi]}$ and $p^{ra}_{[0,\pi]}$, as $e^{-i(y-x)/2}\theta(-e^{i(y-x)}; p_t) = e^{-i(y-x)/2}e^{i(y-x)}\theta(1/(-e^{i(y-x)}); p_t) = e^{-i(x-y)/2}\theta(-e^{i(x-y)}; p_t)$.

2.4 Expressions by Jacobi's Theta Functions

Here we introduce the four kinds of Jacobi's theta functions denoted by ϑ_j, $j = 0, 1, 2, 3$, since they are useful to express the results given above. We use the nome modular parameter τ defined by (2.10). Using the theta function $\theta(z; p)$ defined by (1.6), ϑ_1 is defined as follows [57, 78]:

$$\vartheta_1(\xi; \tau) := ip^{1/8}(p; p)_\infty e^{-i\xi}\theta(e^{2i\xi}; p)$$

$$= i \sum_{n=-\infty}^{\infty} (-1)^n p^{(n-1/2)^2/2} e^{(2n-1)i\xi}$$

$$= 2 \sum_{n=1}^{\infty} (-1)^{n-1} e^{(n-1/2)^2\pi i\tau} \sin\{(2n-1)\xi\}. \quad (2.33)$$

Then other three kinds of Jacobi's theta functions are defined as [57, 78]

$$\vartheta_0(\xi; \tau) := -ie^{i(\xi+\pi\tau/4)}\vartheta_1(\xi + \pi\tau/2; \tau) = (p; p)_\infty \theta(p^{1/2}e^{2i\xi}; p)$$

$$= \sum_{n=-\infty}^{\infty} (-1)^n p^{n^2/2} e^{2\pi i\xi} = 1 + 2 \sum_{n=1}^{\infty} (-1)^n e^{n^2\pi i\tau} \cos(2n\xi), \quad (2.34)$$

$$\vartheta_2(\xi; \tau) := \vartheta_1(\xi + \pi/2; \tau) = p^{1/8}(p; p)_\infty e^{-i\xi}\theta(-e^{2i\xi}; p)$$

$$= \sum_{n=-\infty}^{\infty} p^{(n-1/2)^2/2} e^{(2n-1)i\xi} = 2 \sum_{n=1}^{\infty} e^{(n-1/2)^2\pi i\tau} \cos\{(2n-1)\xi\}, \quad (2.35)$$

$$\vartheta_3(\xi; \tau) := e^{i(\xi+\pi\tau/4)}\vartheta_1(\xi + \pi(1+\tau)/2; \tau) = (p; p)_\infty \theta(-p^{1/2}e^{2i\xi}; p)$$

$$= \sum_{n=-\infty}^{\infty} p^{n^2/2} e^{2\pi i\xi} = 1 + 2 \sum_{n=1}^{\infty} e^{n^2\pi i\tau} \cos(2n\xi). \quad (2.36)$$

Then the transition probability densities of the Brownian motions on $\mathbb{T}$ or in $[0, \pi]$ given in the previous section are expressed as follows [38]: For $t > 0$,

$$p_{\mathbb{T}}(0, x; t, y) = \frac{1}{2\pi}\vartheta_3((y-x)/2; it/2\pi)$$
$$= p(0, x; t, y)\vartheta_3(i\pi(y-x)/t; 2\pi i/t), \quad x, y \in [0, 2\pi), \quad (2.37)$$

and

$$\widetilde{p}_{\mathbb{T}}(0, x; t, y) = \frac{1}{2\pi}\vartheta_2((y-x)/2; it/2\pi), \quad x, y \in [0, 2\pi),$$

$$p_{[0,\pi]}^{aa}(0, x; t, y) = \frac{1}{2\pi}\{\vartheta_3((y-x)/2; it/2\pi) - \vartheta_3((y+x)/2; it/2\pi)\},$$

$$p_{[0,\pi]}^{rr}(0, x; t, y) = \frac{1}{2\pi}\{\vartheta_3((y-x)/2; it/2\pi) + \vartheta_3((y+x)/2; it/2\pi)\},$$

$$p_{[0,\pi]}^{ar}(0, x; t, y) = \frac{1}{2\pi}\{\vartheta_2((y-x)/2; it/2\pi) - \vartheta_2((y+x)/2; it/2\pi)\},$$

$$p_{[0,\pi]}^{ra}(0, x; t, y) = \frac{1}{2\pi}\{\vartheta_2((y-x)/2; it/2\pi) + \vartheta_2((y+x)/2; it/2\pi)\},$$

$$x, y \in [0, \pi]. \quad (2.38)$$

Exercise

2.1 The two expressions of $p_{\mathbb{T}}$ using Jacobi's theta function ϑ_3 in (2.37) correspond to the two expressions in (2.8), respectively. The nome modular parameter is $it/2\pi$ for the first expression, while it is $2\pi i/t$ for the second one, and they are related by Jacobi's imaginary transformation. Give the expressions for other transition probability densities in (2.38) using suitable Jacobi's theta functions with the nome modular parameter $2\pi i/t$.

Chapter 3
Biorthogonal Systems of Theta Functions and Macdonald Denominators

Abstract The fact $\lim_{p \to 0} \theta(\zeta; p) = 1 - \zeta, \zeta \in \mathbb{C}^\times$ suggests that the theta function $\theta(\zeta; p)$ is an elliptic analogue of a linear function of ζ. What is the elliptic analogue of a polynomial of ζ? Rosengren and Schlosser gave seven kinds of answers to this fundamental question by introducing seven infinite series of spaces of theta functions associated with the irreducible reduced affine root systems, $R_n = A_{n-1}, B_n, B_n^\vee, C_n, C_n^\vee, BC_n, n \in \mathbb{N}$, and $D_n, n \in \{2, 3, \dots\}$. Here n indicates the degree of the elliptic analogues of polynomials. The basis functions for the function spaces are called the R_n theta functions and are denoted by $\{\psi_j^{R_n}\}_{j=1}^n$. It was proved that the determinants consisting of $\{\psi_j^{R_n}\}_{j=1}^n$ provide the Macdonald denominator formulas, which are the elliptic extensions of the Weyl denominator formulas. In this chapter, first we give a brief review of the theory of Rosengren and Schlosser. Then we introduce appropriate inner products and prove the biorthogonality relations for the R_n theta functions of Rosengren and Schlosser.

3.1 A_{n-1} Theta Functions and Determinantal Identity

The elliptic analogue of a polynomials of θ might be given by a product of θ's. We have to remark on the equalities (1.12), however, since they show that a degree of product of θ's depends on a choice of nome. In order to define a degree of polynomials of θ's with respect to a specified nome, Rosengren and Schlosser [64] generalized the notion of the quasi-periodicity (1.10) as follows.

Definition 3.1 Assume that $f(\zeta)$ is holomorphic in $\mathbb{C}^\times$ and $n \in \mathbb{N}$. If there is a parameter $r \in \mathbb{R} \setminus \{0\}$ and f satisfies the equality

$$f(p\zeta) = \frac{(-1)^n}{r\zeta^n} f(\zeta),$$

then f is said to be an A_{n-1} *theta function* of *norm* r. The space of all A_{n-1} theta functions with nome p and norm r is denoted by $\mathcal{E}_{p,r}^{A_{n-1}}$.

M. Katori, *Elliptic Extensions in Statistical and Stochastic Systems*,
SpringerBriefs in Mathematical Physics 47,
https://doi.org/10.1007/978-981-19-9527-9_3

By this definition, we can say that $\theta(\zeta; p)$ is an A_0 theta function of norm $r = 1$. The following is proved [74]. (See Exercise 3.1.)

Lemma 3.1 *The space $\mathcal{E}_{p,r}^{A_{n-1}}$ is n-dimensional and $\{\psi_j^{A_{n-1}}(\zeta; p, r)\}_{j=1}^n$ defined by*

$$\psi_j^{A_{n-1}}(\zeta; p, r) := \zeta^{j-1}\theta(p^{j-1}(-1)^{n-1}r\zeta^n; p^n)$$

$$= \zeta^{j-1}\prod_{k=0}^{n-1}\theta(\alpha^{j-1}\beta\omega_n^k\zeta; p), \quad j = 1, \ldots, n, \qquad (3.1)$$

form a basis, where α and β are complex numbers such that $\alpha^n = p$, $\beta^n = (-1)^{n-1}r$, respectively, and ω_n is a primitive n-th root of unity.

By (1.7), we see that

$$\psi_j^{A_{n-1}}(\zeta; 0, r) := \lim_{p \to 0} \psi_j^{A_{n-1}}(\zeta; p, r)$$

$$= \begin{cases} 1 - (-1)^{n-1}r\zeta^n, & j = 1, \\ \zeta^{j-1}, & j = 2, \ldots, n. \end{cases}$$

Hence $\mathcal{E}_{0,r}^{A_{n-1}}$ spanned by them is a space of polynomials of degree n in the form $c_0 + c_1\zeta + \cdots + c_n\zeta^n$ with a constraint such that the ratio c_n/c_0 is fixed to be $-(-1)^{n-1}r$. It implies that $\dim \mathcal{E}_{0,r}^{A_{n-1}} = n$.

It is easy to perform the following calculation:

$$\det_{1 \le j,k \le n}\left[\psi_j^{A_{n-1}}(\zeta_k; 0, r)\right] = \det_{1 \le j,k \le n}\left[\zeta_k^{j-1} - (-1)^{n-1}r\zeta_k^n\delta_{j1}\right]$$

$$= \sum_{\ell=1}^n(-1)^{1+\ell}\left\{1 - (-1)^{n-1}r\zeta_\ell^n\right\}\det_{\substack{1 \le j \le n-1 \\ 1 \le k \le n, k \ne \ell}}\left[\zeta_k^j\right]$$

$$= \det_{1 \le j,k \le n}\left[\zeta_k^{j-1}\right] - r(-1)^{n-1}\det_{1 \le j,k \le n}\left[\zeta_k^{(j-1)+n\delta_{j1}}\right]$$

$$= \left(1 - r\prod_{\ell=1}^n\zeta_\ell\right)\det_{1 \le j,k \le n}\left[\zeta_k^{j-1}\right].$$

Hence we have the equality

$$\det_{1 \le j,k \le n}\left[\psi_j^{A_{n-1}}(\zeta_k; 0, r)\right] = \left(1 - r\prod_{\ell=1}^n\zeta_\ell\right)W^{A_{n-1}}(\boldsymbol{\zeta}), \qquad (3.2)$$

for $\boldsymbol{\zeta} = (\zeta_1, \ldots, \zeta_n) \in \mathbb{C}^n$, where

$$W^{A_{n-1}}(\boldsymbol{\zeta}) := \prod_{1 \le j < k \le n}(\zeta_k - \zeta_j). \qquad (3.3)$$

If we set $r = 0$ as well as $p = 0$, $\mathcal{E}_{0,0}^{A_{n-1}}$ is the space of all polynomials of degree $n - 1$ without any restriction on coefficients. In this case, (3.2) is reduced to

$$W^{A_{n-1}}(\zeta) = \det_{1 \le j,k \le n} \left[\zeta_k^{j-1} \right] = \prod_{1 \le j < k \le n} (\zeta_k - \zeta_j). \tag{3.4}$$

The multivariate polynomial (3.3) is well known as the *Vandermonde determinant*. It gives the *Weyl denominator of type* A_{n-1}, and the determinantal identity (3.4) is called the *Weyl denominator formula of type* A_{n-1}.

The elliptic extension of $W^{A_{n-1}}$ is called the *Macdonald denominator of type* A_{n-1} and defined as follows [54, 64].

Definition 3.2 For $\zeta = (\zeta_1, \ldots, \zeta_n) \in (\mathbb{C}^\times)^n$, the Macdonald denominator of type A_{n-1} is defined as

$$M^{A_{n-1}}(\zeta; p) := \prod_{1 \le j < k \le n} \zeta_k \theta(\zeta_j/\zeta_k; p). \tag{3.5}$$

As we observed above, this is an elliptic extension of the Weyl denominator of type A_{n-1} in the sense,

$$\lim_{p \to 0} M^{A_{n-1}}(\zeta; p) = W^{A_{n-1}}(\zeta). \tag{3.6}$$

The elliptic extension of the determinantal identity (3.2) was proved by Rosengren and Schlosser [64]. Similar identities are found in the papers by Forrester [11, 12, 14]. See also Proposition 5.6.3 on page 216 of the textbook of Forrester [16].

Proposition 3.1 ([11, 64]) *The following determinantal identity holds:*

$$\det_{1 \le j,k \le n} \left[\psi_j^{A_{n-1}}(\zeta_k; p, r) \right] = \frac{(p; p)_\infty^n}{(p^n; p^n)_\infty^n} \theta \left(r \prod_{\ell=1}^n \zeta_\ell; p \right) M^{A_{n-1}}(\zeta; p). \tag{3.7}$$

(See Exercise 3.2.)

3.2 Other R_n Theta Functions and Determinantal Identities

There are seven infinite families of *irreducible reduced affine root systems*, which are denoted by A, B, $B^\vee$, C, $C^\vee$, BC and D [10, 54]. Associated with them, Rosengren and Schlosser [64] defined seven types of theta functions. One of them is the A_{n-1} type of theta functions, which was already defined by Definition 3.1. The others are defined as follows. From now on we assume $n \in \mathbb{N}$ for $R_n = B_n$, $B_n^\vee$, C_n, $C_n^\vee$, BC_n, and $n \in \mathbb{N} \setminus \{1\} = \{2, 3, \ldots\}$ for $R_n = D_n$.

Definition 3.3 Assume that $f(\zeta)$ is holomorphic in $\mathbb{C}^\times$. For $R_n = B_n,\, B_n^\vee,\, C_n,\, C_n^\vee,$ $BC_n,\, D_n$, we call f an R_n *theta function* if the following are satisfied:

$$f(p\zeta) = -\frac{1}{p^{n-1}\zeta^{2n-1}}f(\zeta), \quad f(1/\zeta) = -\frac{1}{\zeta}f(\zeta), \quad R_n = B_n,$$

$$f(p\zeta) = -\frac{1}{p^n\zeta^{2n}}f(\zeta), \quad f(1/\zeta) = -f(\zeta), \quad R_n = B_n^\vee,$$

$$f(p\zeta) = \frac{1}{p^{n+1}\zeta^{2n+2}}f(\zeta), \quad f(1/\zeta) = -f(\zeta), \quad R_n = C_n,$$

$$f(p\zeta) = \frac{1}{p^{n-1/2}\zeta^{2n}}f(\zeta), \quad f(1/\zeta) = -\frac{1}{\zeta}f(\zeta), \quad R_n = C_n^\vee,$$

$$f(p\zeta) = \frac{1}{p^n\zeta^{2n+1}}f(\zeta), \quad f(1/\zeta) = -\frac{1}{\zeta}f(\zeta), \quad R_n = BC_n,$$

$$f(p\zeta) = \frac{1}{p^{n-1}\zeta^{2n-2}}f(\zeta), \quad f(1/\zeta) = f(\zeta), \quad R_n = D_n. \tag{3.8}$$

The space of all R_n theta functions with nome p is denoted by $\mathcal{E}_p^{R_n}$.

In order to clarify the common structure of (3.8), we introduce the following notations:

$$N = N^{R_n} := \begin{cases} 2n-1, & R_n = B_n, \\ 2n, & R_n = B_n^\vee, C_n^\vee, \\ 2n+2, & R_n = C_n, \\ 2n+1, & R_n = BC_n, \\ 2n-2, & R_n = D_n, \end{cases} \tag{3.9}$$

$$a = a^{R_n} := \begin{cases} 1, & R_n = B_n, C_n^\vee, BC_n, \\ 0, & R_n = B_n^\vee, C_n, D_n, \end{cases} \tag{3.10}$$

$$\sigma_1 = \sigma_1^{R_n} := \begin{cases} -1, & R_n = B_n, B_n^\vee, \\ 1, & R_n = C_n, C_n^\vee, BC_n, D_n. \end{cases} \tag{3.11}$$

$$\sigma_2 = \sigma_2^{R_n} := \begin{cases} -1, & R_n = B_n, B_n^\vee, C_n, C_n^\vee, BC_n, \\ 1, & R_n = D_n. \end{cases} \tag{3.12}$$

Then (3.8) is expressed as

$$f(p\zeta) = \sigma_1 \frac{1}{p^{(N-a)/2}\zeta^N}f(\zeta), \tag{3.13}$$

$$f(1/\zeta) = \sigma_2 \frac{1}{\zeta^a}f(\zeta). \tag{3.14}$$

In addition to (3.9)–(3.12), for $j = 1, \ldots, n$, we put

$$\alpha_j = \alpha_j^{R_n} := \begin{cases} j - n, & R_n = B_n, C_n^\vee, BC_n, D_n, \\ j - n - 1, & R_n = B_n^\vee, C_n, \end{cases} \tag{3.15}$$

$$\beta_j(p) = \beta_j^{R_n}(p) := -\sigma_1 p^{\alpha_j + (N-a)/2} = \begin{cases} p^{j-1}, & R_n = B_n, B_n^\vee, \\ -p^j, & R_n = C_n, BC_n, \\ -p^{j-1/2}, & R_n = C_n^\vee, \\ -p^{j-1}, & R_n = D_n. \end{cases} \tag{3.16}$$

The following was proved [64].

Lemma 3.2 *For $R_n = B_n, B_n^\vee, C_n, C_n^\vee, BC_n, D_n$, the space $\mathcal{E}_p^{R_n}$ is n-dimensional and*

$$\psi_j^{R_n}(\zeta; p) = \zeta^{\alpha_j} \theta(\beta_j(p)\zeta^N; p^N) + \sigma_2 \zeta^{-(\alpha_j - a)} \theta(\beta_j(p)\zeta^{-N}; p^N), \tag{3.17}$$

$j = 1, \ldots, n$, form a basis.

For convenience, the explicit expressions of (3.17) are given below:

$$\psi_j^{B_n}(\zeta; p) := \zeta^{j-n} \theta(p^{j-1}\zeta^{2n-1}; p^{2n-1}) - \zeta^{n+1-j} \theta(p^{j-1}\zeta^{1-2n}; p^{2n-1}),$$
$$\psi_j^{B_n^\vee}(\zeta; p) := \zeta^{j-n-1} \theta(p^{j-1}\zeta^{2n}; p^{2n}) - \zeta^{n+1-j} \theta(p^{j-1}\zeta^{-2n}; p^{2n}),$$
$$\psi_j^{C_n}(\zeta; p) := \zeta^{j-n-1} \theta(-p^j\zeta^{2n+2}; p^{2n+2}) - \zeta^{n+1-j} \theta(-p^j\zeta^{-2n-2}; p^{2n+2}),$$
$$\psi_j^{C_n^\vee}(\zeta; p) := \zeta^{j-n} \theta(-p^{j-1/2}\zeta^{2n}; p^{2n}) - \zeta^{n+1-j} \theta(-p^{j-1/2}\zeta^{-2n}; p^{2n}),$$
$$\psi_j^{BC_n}(\zeta; p) := \zeta^{j-n} \theta(-p^j\zeta^{2n+1}; p^{2n+1}) - \zeta^{n+1-j} \theta(-p^j\zeta^{-2n-1}; p^{2n+1}),$$
$$\psi_j^{D_n}(\zeta; p) := \zeta^{j-n} \theta(-p^{j-1}\zeta^{2n-2}; p^{2n-2}) + \zeta^{n-j} \theta(-p^{j-1}\zeta^{-2n+2}; p^{2n-2}). \tag{3.18}$$

(See Exercises 3.3 and 3.4.)

In addition to the Macdonald denominator of type A_{n-1} given by Definition 3.2, the other six kinds of Macdonald denominators are defined as follows [64].

Definition 3.4 For $\boldsymbol{\zeta} = (\zeta_1, \ldots, \zeta_n) \in (\mathbb{C}^\times)^n$, let

$$M^{B_n}(\boldsymbol{\zeta}; p) := \prod_{\ell=1}^n \theta(\zeta_\ell; p) \prod_{1 \le j < k \le n} \zeta_j^{-1} \theta(\zeta_j \zeta_k; p) \theta(\zeta_j/\zeta_k; p),$$

$$M^{B_n^\vee}(\boldsymbol{\zeta}; p) := \prod_{\ell=1}^n \zeta_\ell^{-1} \theta(\zeta_\ell^2; p^2) \prod_{1 \le j < k \le n} \zeta_j^{-1} \theta(\zeta_j \zeta_k; p) \theta(\zeta_j/\zeta_k; p),$$

$$M^{C_n}(\boldsymbol{\zeta}; p) := \prod_{\ell=1}^{n} \zeta_\ell^{-1} \theta(\zeta_\ell^2; p) \prod_{1 \le j < k \le n} \zeta_j^{-1} \theta(\zeta_j \zeta_k; p) \theta(\zeta_j/\zeta_k; p),$$

$$M^{C_n^\vee}(\boldsymbol{\zeta}; p) := \prod_{\ell=1}^{n} \theta(\zeta_\ell; p^{1/2}) \prod_{1 \le j < k \le n} \zeta_j^{-1} \theta(\zeta_j \zeta_k; p) \theta(\zeta_j/\zeta_k; p),$$

$$M^{BC_n}(\boldsymbol{\zeta}; p) := \prod_{\ell=1}^{n} \theta(\zeta_\ell; p) \theta(p\zeta_\ell^2; p^2) \prod_{1 \le j < k \le n} \zeta_j^{-1} \theta(\zeta_j \zeta_k; p) \theta(\zeta_j/\zeta_k; p),$$

$$M^{D_n}(\boldsymbol{\zeta}; p) := \prod_{1 \le j < k \le n} \zeta_j^{-1} \theta(\zeta_j \zeta_k; p) \theta(\zeta_j/\zeta_k; p). \tag{3.19}$$

They are called the *Macdonald denominators of type* R_n for $R_n = B_n, B_n^\vee, C_n, C_n^\vee$, BC_n, D_n, respectively.

The above are regarded as the elliptic extensions of the *Weyl denominators of types* R_n with $R_n = B_n, C_n$ and D_n:

$$W^{B_n}(\boldsymbol{\zeta}) = \det_{1 \le j,k \le n} \left[\zeta_k^{j-n} - \zeta_k^{n+1-j} \right] = \prod_{\ell=1}^{n} \zeta_\ell^{1-n} (1 - \zeta_\ell) \prod_{1 \le j < k \le n} (\zeta_k - \zeta_j)(1 - \zeta_j \zeta_k)$$

$$= \lim_{p \to 0} M^{B_n}(\boldsymbol{\zeta}; p) = \lim_{p \to 0} M^{C_n^\vee}(\boldsymbol{\zeta}; p) = \lim_{p \to 0} M^{BC_n}(\boldsymbol{\zeta}; p),$$

$$W^{C_n}(\boldsymbol{\zeta}) = \det_{1 \le j,k \le n} \left[\zeta_k^{j-n-1} - \zeta_k^{n+1-j} \right] = \prod_{\ell=1}^{n} \zeta_\ell^{-n} (1 - \zeta_\ell^2) \prod_{1 \le j < k \le n} (\zeta_k - \zeta_j)(1 - \zeta_j \zeta_k)$$

$$= \lim_{p \to 0} M^{B_n^\vee}(\boldsymbol{\zeta}; p) = \lim_{p \to 0} M^{C_n}(\boldsymbol{\zeta}; p),$$

$$W^{D_n}(\boldsymbol{\zeta}) = \det_{1 \le j,k \le n} \left[\zeta_k^{j-n} + \zeta_k^{n-j} \right] = 2 \prod_{\ell=1}^{n} \zeta_\ell^{1-n} \prod_{1 \le j < k \le n} (\zeta_k - \zeta_j)(1 - \zeta_j \zeta_k)$$

$$= 2 \lim_{p \to 0} M^{D_n}(\boldsymbol{\zeta}; p). \tag{3.20}$$

Rosengren and Schlosser proved the following *Macdonald denominator formulas of type* R_n for $R_n = B_n, B_n^\vee, C_n, C_n^\vee, BC_n$, and D_n [64].[1]

Proposition 3.2 ([64]) *The following equalities hold for* $\boldsymbol{\zeta} = (\zeta_1, \ldots, \zeta_n) \in (\mathbb{C}^\times)^n$:

$$\det_{1 \le j,k \le n} \left[\psi_j^{B_n}(\zeta_k; p) \right] = \frac{2(p; p)_\infty^n}{(p^{2n-1}; p^{2n-1})_\infty^n} M^{B_n}(\boldsymbol{\zeta}; p),$$

$$\det_{1 \le j,k \le n} \left[\psi_j^{B_n^\vee}(\zeta_k; p) \right] = \frac{2(p^2; p^2)_\infty (p; p)_\infty^{n-1}}{(p^{2n}; p^{2n})_\infty^n} M^{B_n^\vee}(\boldsymbol{\zeta}; p),$$

---

[1] Other elliptic determinantal formulas can be found in the papers by Frobenius [18, 19]. In the context of integrable systems [65], Hasegawa reported some elliptic determinant evaluations in [28, Lemma 1]. See also [30, 51, 62, 63, 66], [74, Appendix B], [75], [77, Theorem 4.17, Lemma 5.3].

$$\det_{1 \le j,k \le n} \left[\psi_j^{C_n}(\zeta_k; p) \right] = \frac{(p; p)_\infty^n}{(p^{2n+2}; p^{2n+2})_\infty^n} M^{C_n}(\boldsymbol{\zeta}; p),$$

$$\det_{1 \le j,k \le n} \left[\psi_j^{C_n^\vee}(\zeta_k; p) \right] = \frac{(p^{1/2}; p^{1/2})_\infty (p; p)_\infty^{n-1}}{(p^{2n}, p^{2n})_\infty^n} M^{C_n^\vee}(\boldsymbol{\zeta}; p),$$

$$\det_{1 \le j,k \le n} \left[\psi_j^{BC_n}(\zeta_k; p) \right] = \frac{(p; p)_\infty^n}{(p^{2n+1}; p^{2n+1})_\infty^n} M^{BC_n}(\boldsymbol{\zeta}; p),$$

$$\det_{1 \le j,k \le n} \left[\psi_j^{D_n}(\zeta_k; p) \right] = \frac{4(p; p)_\infty^n}{(p^{2n-2}; p^{2n-2})_\infty^n} M^{D_n}(\boldsymbol{\zeta}; p). \tag{3.21}$$

3.3 Biorthogonality of A_{n-1} Theta Functions

For a pair of spaces $(\mathcal{E}_{p,r}^{A_{n-1}}, \mathcal{E}_{\widehat{p},\widehat{r}}^{A_{n-1}})$ of the A_{n-1} theta functions, $p, \widehat{p} \in (0, 1), r, \widehat{r} \in \mathbb{R} \setminus \{0\}$, we introduce the following inner product:

$$\langle f, g \rangle_{\mathbb{T}} := \frac{1}{2\pi} \int_{|\zeta|=1} f(\zeta)\overline{g(\zeta)}\ell(d\zeta)$$

$$= \frac{1}{2\pi} \int_0^{2\pi} f(e^{ix})\overline{g(e^{ix})}dx, \quad f \in \mathcal{E}_{p,r}^{A_{n-1}}, \ g \in \mathcal{E}_{\widehat{p},\widehat{r}}^{A_{n-1}}, \tag{3.22}$$

where ℓ denotes the arc length measure on the circle $\mathbb{T}$ normalized as $\ell(\mathbb{T}) = 2\pi$. Let $\mathbf{1}(\omega)$ be an indicator of the condition ω such that $\mathbf{1}(\omega) = 1$ if ω is satisfied and $\mathbf{1}(\omega) = 0$ otherwise. For the special case $\omega = \{i = j\}, i, j \in \mathbb{Z}$, the indicator is denoted by Kronecker's delta; $\delta_{ij} := \mathbf{1}(i = j)$.

Then we can prove the following [38].

Proposition 3.3 *Let $p, \widehat{p} \in (0, 1)$ and $r, \widehat{r} \in \mathbb{R} \setminus \{0\}$. Then*

$$\langle \psi_j^{A_{n-1}}(\cdot; p, r), \psi_k^{A_{n-1}}(\cdot; \widehat{p}, \widehat{r}) \rangle_{\mathbb{T}} = h_j^{A_{n-1}}(p, \widehat{p}, r\widehat{r})\delta_{jk}, \tag{3.23}$$

for $j, k = 1, \ldots, n$, where

$$h_j^{A_{n-1}}(p, \widehat{p}, r\widehat{r}) := \frac{((p\widehat{p})^n; (p\widehat{p})^n)_\infty}{(p^n; p^n)_\infty (\widehat{p}^n, \widehat{p}^n)_\infty} \theta(-(r\widehat{r})(p\widehat{p})^{j-1}; (p\widehat{p})^n). \tag{3.24}$$

Proof We apply the Laurent expansion of the theta function (1.13) to the definition (3.1). Then we have

$$\psi_j^{A_{n-1}}(e^{ix}; p, r) = \frac{e^{i(j-1)x}}{(p^n; p^n)_\infty} \sum_{\ell \in \mathbb{Z}} (-1)^\ell p^{n\binom{\ell}{2}} (-1)^{(n-1)\ell} r^\ell p^{(j-1)\ell} e^{in\ell x},$$

and

$$\overline{\psi_k^{A_{n-1}}(e^{ix};\widehat{p},\widehat{r})} = \frac{e^{-i(k-1)x}}{(\widehat{p}^n;\widehat{p}^n)_\infty} \sum_{m\in\mathbb{Z}} (-1)^m \widehat{p}^{n\binom{m}{2}}(-1)^{(n-1)m}\widehat{r}^m \widehat{p}^{(k-1)m} e^{-inmx}.$$

Hence we see that the inner product of them includes the following integrals:

$$I_{jk,\ell m}^{A_{n-1}} := \frac{1}{2\pi}\int_0^{2\pi} e^{i\{(j-k)+n(\ell-m)\}x}dx$$

as

$$\langle \psi_j^{A_{n-1}}(\cdot;p,r), \psi_k^{A_{n-1}}(\cdot;\widehat{p},\widehat{r})\rangle_{\mathbb{T}}$$
$$= \frac{1}{(p^n;p^n)_\infty(\widehat{p}^n;\widehat{p}^n)_\infty} \sum_{\ell\in\mathbb{Z}}\sum_{m\in\mathbb{Z}} (-1)^{n(\ell+m)} p^{n\binom{\ell}{2}}\widehat{p}^{n\binom{m}{2}} r^\ell \widehat{r}^m p^{(j-1)\ell}\widehat{p}^{(k-1)m} I_{jk,\ell m}^{A_{n-1}}.$$

It is easy to verify that $I_{jk,\ell m}^{A_{n-1}} = \mathbf{1}((j-k)+n(\ell-m)=0)$. Since $\ell, m \in \mathbb{Z}$, while $j,k \in \{1,\ldots,n\}$, $|j-k| \le n-1 < n$ and thus the nonzero condition of $I_{jk,\ell m}^{A_{n-1}}$ is satisfied if and only if $j-k=0$ and $\ell-m=0$. That is, we have the equalities

$$I_{jk,\ell m}^{A_{n-1}} = \delta_{jk}\delta_{\ell m} \quad \text{for } j,k \in \{1,\ldots,n\}, \ \ell, m \in \mathbb{Z}. \tag{3.25}$$

Therefore, we obtain

$$\langle \psi_j^{A_{n-1}}(\cdot;p,r), \psi_k^{A_{n-1}}(\cdot;\widehat{p},\widehat{r})\rangle_{\mathbb{T}}$$
$$= \frac{\delta_{jk}}{(p^n;p^n)_\infty(\widehat{p}^n;\widehat{p}^n)_\infty} \sum_{\ell\in\mathbb{Z}} (p\widehat{p})^{n\binom{\ell}{2}}(r\widehat{r})^\ell (p\widehat{p})^{(j-1)\ell}.$$

Again we use the Laurent expansion (1.13) of the theta function and then (3.23) with (3.24) is concluded. □

When $p \ne \widehat{p}$, $r \ne \widehat{r}$, we have two distinct sets of functions $\{\psi_j^{A_{n-1}}(\cdot;p,r)\}_{j=1}^n$ and $\{\psi_j^{A_{n-1}}(\cdot;\widehat{p},\widehat{r})\}_{j=1}^n$. In such a general case, the property asserted by Proposition 3.3 is called *biorthononality* [38]. As a special case with $p = \widehat{p}$ and $r = \widehat{r}$, $\{\psi_j^{A_{n-1}}(\cdot;p,r)\}_{j=1}^n$ makes an orthogonal basis of $\mathcal{E}_{p,r}^{A_{n-1}}$ (see pp. 216–217 of [16]).

3.4 Biorthogonality of Other R_n Theta Functions

For a pair of spaces of other theta functions $(\mathcal{E}_p^{R_n}, \mathcal{E}_{\widehat{p}}^{R_n})$, $p, \widehat{p} \in (0,1)$, for $R_n = B_n$, $B_n^\vee$, C_n, $C_n^\vee$, BC_n, D_n, the inner product is introduced as

$$\langle f, g \rangle_{[0,\pi]} := \frac{1}{\pi} \int_0^\pi f(e^{ix}) \overline{g(e^{ix})} dx, \quad f \in \mathcal{E}_p^{R_n}, \quad g \in \mathcal{E}_{\widehat{p}}^{R_n}. \tag{3.26}$$

The biorthogonality can be also proved as follows.

Proposition 3.4 *Let $p, \widehat{p} \in (0, 1)$. Then for $R_n = B_n, B_n^\vee, C_n, C_n^\vee, BC_n, D_n$,*

$$\langle \psi_j^{R_n}(\cdot; p), \psi_k^{R_n}(\cdot; \widehat{p}) \rangle_{[0,\pi]} = h_j^{R_n}(p, \widehat{p}) \delta_{jk}, \quad j, k = 1, \ldots, n, \tag{3.27}$$

with

$$h_j^{R_n}(p, \widehat{p}) := c_j \frac{2((p\widehat{p})^N; (p\widehat{p})^N)_\infty}{(p^N; p^N)_\infty (\widehat{p}^N; \widehat{p}^N)_\infty} \theta(-\beta_j(p)\beta_j(\widehat{p}); (p\widehat{p})^N), \tag{3.28}$$

where the factors are given by

$$c_j = c_j^{R_n} := \begin{cases} 1, & j = 1, \ldots, n, \quad R_n = C_n, C_n^\vee, BC_n, \\ \begin{cases} 2, & j = 1, \\ 1, & j = 2, \ldots, n, \end{cases} & R_n = B_n, B_n^\vee, \\ \begin{cases} 2, & j = 1, n, \\ 1, & j = 2, \ldots, n-1, \end{cases} & R_n = D_n. \end{cases} \tag{3.29}$$

Here $\beta_j(p) = \beta_j^{R_n}(p)$ and $N = N^{R_n}$ are given by (3.16) and (3.9) for each R_n.

Proof Using the expressions (3.17) with (3.9)–(3.12), (3.15) and (3.16), we have

$$\langle \psi_j^{R_n}(\cdot; p), \psi_k^{R_n}(\cdot; \widehat{p}) \rangle_{[0,\pi]}$$

$$= \frac{1}{\pi} \int_0^\pi \left\{ e^{i\alpha_j x} \theta(\beta_j(p)e^{iNx}; p^N) + \sigma_2 e^{-i(\alpha_j - a)x} \theta(\beta_j(p)e^{-iNx}; p^N) \right\}$$

$$\times \left\{ e^{-i\alpha_k x} \theta(\beta_k(\widehat{p})e^{-iNx}; p^N) + \sigma_2 e^{i(\alpha_k - a)x} \theta(\beta_k(\widehat{p})e^{iNx}; p^N) \right\} dx$$

$$= \frac{1}{\pi} \int_0^\pi e^{i(j-k)x} \theta(\beta_j(p)e^{iNx}; p^N) \theta(\beta_k(\widehat{p})e^{-iNx}; \widehat{p}^N) dx$$

$$+ \frac{1}{\pi} \int_0^\pi dx \, e^{-i(j-k)x} \theta(\beta_j(p)e^{-iNx}; p^N) \theta(\beta_k(\widehat{p})e^{iNx}; \widehat{p}^N) dx$$

$$+ \sigma_2 \frac{1}{\pi} \int_0^\pi dx \, e^{i(j+k-\delta)x} \theta(\beta_j(p)e^{iNx}; p^N) \theta(\beta_k(\widehat{p})e^{iNx}; \widehat{p}^N) dx$$

$$+ \sigma_2 \frac{1}{\pi} \int_0^\pi dx \, e^{-i(j+k-\delta)x} \theta(\beta_j(p)e^{-iNx}; p^N) \theta(\beta_k(\widehat{p})e^{-iNx}; \widehat{p}^N) dx, \tag{3.30}$$

where we have used the facts concluded from (3.15) such that

$$\alpha_j - \alpha_k = j - k,$$
$$\alpha_j + \alpha_k - a = j + k - \delta \tag{3.31}$$

with

$$
\delta = \delta^{R_n} := \begin{cases} 2n+1, & R_n = B_n, C_n^\vee, BC_n, \\ 2n+2, & R_n = B_n^\vee, C_n, \\ 2n, & R_n = D_n. \end{cases} \tag{3.32}
$$

By changing the integral variable in the third and fourth terms in the rightmost side of (3.30) as $x \to -x$, the above is written as

$$
\langle \psi_j^{R_n}(\cdot; p), \psi_k^{R_n}(\cdot; \widehat{p}) \rangle_{[0,\pi]} = 2(J_{jk,+}^{R_n} + \sigma_2^{R_n} J_{jk,-}^{R_n}), \tag{3.33}
$$

with

$$
J_{jk,+}^{R_n} := \frac{1}{2\pi} \int_{-\pi}^{\pi} e^{i(j-k)x} \theta(\beta_j(p)e^{iNx}; p^N) \theta(\beta_k(\widehat{p})e^{-iNx}; \widehat{p}^N) dx,
$$

$$
J_{jk,-}^{R_n} := \frac{1}{2\pi} \int_{-\pi}^{\pi} e^{i(j+k-\delta)x} \theta(\beta_j(p)e^{iNx}; p^N) \theta(\beta_k(\widehat{p})e^{iNx}; \widehat{p}^N) dx. \tag{3.34}
$$

Using the Laurent expansion (1.13), we obtain

$$
J_{jk,\pm}^{R_n} = \frac{1}{(p^N; p^N)_\infty (\widehat{p}^N; \widehat{p}^N)} \sum_{\ell \in \mathbb{Z}} \sum_{m \in \mathbb{Z}} (-1)^{\ell+m} p^{N\binom{\ell}{2}} \widehat{p}^{N\binom{m}{2}} \beta_j(p)^\ell \beta_k(\widehat{p})^m I_{jk,\ell m,\pm}^{R_n},
$$

with

$$
I_{jk,\ell m,+}^{R_n} := \frac{1}{2\pi} \int_{-\pi}^{\pi} e^{i\{(j-k)+N(\ell-m)\}x} dx = \mathbf{1}((j-k) + N(\ell-m) = 0),
$$

$$
I_{jk,\ell m,-}^{R_n} := \frac{1}{2\pi} \int_{-\pi}^{\pi} e^{i\{(j+k-\delta)+N(\ell+m)\}x} dx = \mathbf{1}((j+k-\delta) + N(\ell+m) = 0).
$$

Since $j, k \in \{1, \ldots, n\}$ and hence $|j - k| \le n - 1 < N$ for any R_n, $I_{jk,\ell m,+}^{R_n} = \delta_{jk}\delta_{\ell m}$. Therefore, we have

$$
J_{jk,+}^{R_n} = \frac{\delta_{jk}}{(p^N; p^N)_\infty (\widehat{p}^N; \widehat{p}^N)_\infty} \sum_{\ell \in \mathbb{Z}} (p\widehat{p})^{N\binom{\ell}{2}} (\beta_j(p)\beta_j(\widehat{p}))^\ell
$$

$$
= \frac{\delta_{jk}((p\widehat{p})^N; (p\widehat{p})^N)_\infty}{(p^N; p^N)_\infty (\widehat{p}^N; \widehat{p}^N)_\infty} \theta(-\beta_j(p)\beta_j(\widehat{p}); (p\widehat{p})^N). \tag{3.35}
$$

If $R_n = C_n, C_n^\vee, BC_n$, then we have the inequalities

$$
-N < j + k - \delta < 0.
$$

In this case the condition $(j + k - \delta) + \mathcal{N}(\ell + m) = 0$ cannot be satisfied by any $j, k \in \{1, \ldots, n\}$, $\ell, m \in \mathbb{Z}$, and hence $I^{R_n}_{jk,\ell m,-} = 0$ and then

$$J^{R_n}_{jk,-} = 0, \quad R_n = C_n, C_n^{\vee}, BC_n. \tag{3.36}$$

If $R_n = B_n, B_n^{\vee}$, we have

$$-\mathcal{N} \le j + k - \delta < 0,$$

in which the equality holds if and only if $j = k = 1$. This implies

$$I^{R_n}_{jk,\ell m,-} = \delta_{j1}\delta_{k1}\mathbf{1}(m = -\ell + 1), \quad R_n = B_n, B_n^{\vee}.$$

We notice that $\binom{-\ell+1}{2} = (-\ell+1)(-\ell)/2 = \binom{\ell}{2}$ and $\beta_1(p) = 1$ for $R_n = B_n, B_n^{\vee}$. We obtain

$$
\begin{aligned}
J^{R_n}_{jk,-} &= -\frac{\delta_{j1}\delta_{k1}}{(p^{\mathcal{N}}; p^{\mathcal{N}})_\infty (\widehat{p}^{\mathcal{N}}; \widehat{p}^{\mathcal{N}})_\infty} \sum_{\ell \in \mathbb{Z}} (p\widehat{p})^{\mathcal{N}\binom{\ell}{2}} \\
&= -\frac{\delta_{j1}\delta_{k1}((p\widehat{p})^{\mathcal{N}}; (p\widehat{p})^{\mathcal{N}})_\infty}{(p^{\mathcal{N}}; p^{\mathcal{N}})_\infty (\widehat{p}^{\mathcal{N}}; \widehat{p}^{\mathcal{N}})_\infty} \theta(-1; (p\widehat{p})^{\mathcal{N}}), \quad R_n = B_n, B_n^{\vee}. \tag{3.37}
\end{aligned}
$$

If $R_n = D_n$, we have

$$-\mathcal{N} \le j + k - \delta \le 0,$$

in which the left equality holds if and only if $j = k = 1$, and the right equality holds if and only if $j = k = n$, respectively. Hence

$$I^{R_n}_{jk,\ell m,-} = \delta_{j1}\delta_{k1}\mathbf{1}(m = -\ell + 1) + \delta_{jn}\delta_{kn}\mathbf{1}(m = -\ell).$$

Note that $\binom{-\ell}{2} = (-\ell)(-\ell-1)/2 = \binom{\ell}{2} + \ell$, $\beta_1(p) = -1$, $\beta_n(p) = -p^{n-1}$, and $\mathcal{N} = 2n - 2$ for $R_n = D_n$. Then we obtain

$$
\begin{aligned}
J^{D_n}_{jk,-} = &\frac{1}{(p^{\mathcal{N}}; p^{\mathcal{N}})_\infty (\widehat{p}^{\mathcal{N}}; \widehat{p}^{\mathcal{N}})_\infty} \\
&\times \left\{ \delta_{j1}\delta_{k1} \sum_{\ell \in \mathbb{Z}} (p\widehat{p})^{\mathcal{N}\binom{\ell}{2}} + \delta_{jn}\delta_{kn} \sum_{\ell \in \mathbb{Z}} (p\widehat{p})^{\mathcal{N}\binom{\ell}{2}} \widehat{p}^{\mathcal{N}\ell} (-p^{n-1})^{\ell} (-\widehat{p}^{n-1})^{-\ell} \right\}.
\end{aligned}
$$

Since $\mathcal{N}^{D_n} = 2(n-1)$, we see $\widetilde{p}^{\mathcal{N}\ell}(-\widehat{p}^{n-1})^{-\ell} = (-\widehat{p}^{n-1})^{\ell}$. Hence we have

$$
\begin{aligned}
J^{D_n}_{jk,-} = &\frac{((p\widehat{p})^{\mathcal{N}}; (p\widehat{p})^{\mathcal{N}})_\infty}{(p^{\mathcal{N}}; p^{\mathcal{N}})_\infty (\widehat{p}^{\mathcal{N}}; \widehat{p}^{\mathcal{N}})_\infty} \\
&\times \left\{ \delta_{j1}\delta_{k1}\theta(-1; (p\widehat{p})^{\mathcal{N}}) + \delta_{jn}\delta_{kn}\theta(-(p\widehat{p})^{n-1}; (p\widehat{p})^{\mathcal{N}}) \right\}. \tag{3.38}
\end{aligned}
$$

Inserting (3.35), (3.36), (3.37) and (3.38) into (3.33), the assertion is proved. □

Exercises

3.1 Prove the following equalities:

$$\psi_j^{A_{n-1}}(\omega_n\zeta; p, r) = \omega_n^{j-1}\psi_j^{A_{n-1}}(\zeta; p, r), \quad j = 1, \ldots, n, \tag{3.39}$$

$$\psi_j^{A_{n-1}}(p^{1/n}\zeta; p, r) = p^{(j-1)/n}\zeta^{-1}\psi_{j+1}^{A_{n-1}}(\zeta; p, r), \quad j = 1, \ldots, n-1, \tag{3.40}$$

and

$$\psi_n^{A_{n-1}}(p^{1/n}\zeta; p, r) = -p^{(n-1)/n}\zeta^{-1}(-1)^{n-1}r^{-1}\psi_1^{A_{n-1}}(\zeta; p, r), \tag{3.41}$$

where ω_n in (3.39) is a primitive n-th root of unity.

3.2 The Macdonald denominator formula of A_{n-1} given by (3.7) in Proposition 3.1 is given as follows for $n = 2$:

$$\det\begin{bmatrix} \psi_1^{A_1}(\zeta_1; p, r) & \psi_1^{A_1}(\zeta_2; p, r) \\ \psi_2^{A_1}(\zeta_1; p, r) & \psi_2^{A_1}(\zeta_2; p, r) \end{bmatrix} = \frac{(p; p)_\infty^2}{(p^2; p^2)_\infty^2}\theta(r\zeta_1\zeta_2; p)M^{A_1}(\zeta_1, \zeta_2), \tag{3.42}$$

where

$$\psi_1^{A_1}(\zeta; p, r) = \theta(-r\zeta^2; p^2), \quad \psi_2^{A_1}(\zeta; p, r) = \zeta\theta(-pr\zeta^2; p^2),$$
$$M^{A_1}(\zeta_1, \zeta_2) = \zeta_2\theta(\zeta_1/\zeta_2; p).$$

Prove the equality (3.42) by direct calculation as follows. Let $L(\zeta_1, \zeta_2)$ be the LHS of (3.42) and $R(\zeta_1, \zeta_2) := \theta(r\zeta_1\zeta_2; p)M^{A_1}(\zeta_1, \zeta_2)$.

(1) Since $\psi_j^{A_1}(\zeta; p, r) \in \mathcal{E}_{p,r}^{A_1}$, $j = 1, 2$, $L(\zeta_1, \zeta_2)$ is the A_1 theta function of nome p and norm r with respect to both of ζ_1 and ζ_2. Let $f_1(\zeta) := R(\zeta, \zeta_2)$ with a given $\zeta_2 \in \mathbb{C}^\times$. Prove that $f_1(\zeta)$ is an A_1 theta function of ζ. Similarly, let $f_2(\zeta) := R(\zeta_1, \zeta)$ with a given $\zeta_1 \in \mathbb{C}^\times$, and prove that $f_2(\zeta) \in \mathcal{E}_{p,r}^{A_1}$.

(2) Since $\theta(1; p) = 0$ by definition, it is obvious that $R(\zeta_1, \zeta_2) = 0$, if $\zeta_1 = \zeta_2$, or $\zeta_1 = 1/r\zeta_2$, or $\zeta_2 = 1/r\zeta_1$. Prove that if $\zeta_1 = \zeta_2$, then $L(\zeta_1, \zeta_2) = 0$, and that if $\zeta_1 = 1/r\zeta_2$ or $\zeta_2 = 1/r\zeta_1$, then $L(\zeta_1, \zeta_2) = 0$.

(3) The above results imply that the ratio $L(\zeta_1, \zeta_2)/R(\zeta_1, \zeta_2)$ is a constant c which is independent of ζ_1, ζ_2. Consider the case with $\zeta_1 = 1$ and $\zeta_2 = -1$. Then prove that $c = (p; p)_\infty^2/(p^2; p^2)_\infty^2$.

3.3 Prove the following.

(i) If $R_n = B_n^\vee, C_n, D_n$,

$$\psi_j^{R_n}(-\zeta; p) = (-1)^{\alpha_j}\psi_j^{R_n}(\zeta; p), \quad j = 1, \ldots, n. \tag{3.43}$$

(ii) If $R_n = C_n, C_n^\vee, D_n$,

$$\psi_j^{R_n}(p^{1/2}\zeta; p) = \sigma_2 \zeta^{-N/2} p^{\alpha_j/2} \psi_{n+1-j}^{R_n}(\zeta; p), \quad j = 1, \ldots, n. \qquad (3.44)$$

3.4 Prove that the R_n-theta functions have the following zeros: for $j = 1, \ldots, n$,

$$\psi_j^{R_n}(1; p) = 0, \quad R_n = B_n, B_n^\vee, C_n, C_n^\vee, BC_n, \qquad (3.45)$$

$$\psi_j^{R_n}(-1; p) = 0, \quad R_n = B_n^\vee, C_n, \qquad (3.46)$$

$$\psi_j^{R_n}(p^{1/2}; p) = 0, \quad R_n = C_n, C_n^\vee, BC_n, \qquad (3.47)$$

$$\psi^{R_n}(-p^{1/2}; p) = 0, \quad R_n = C_n, BC_n. \qquad (3.48)$$

Chapter 4
KMLGV Determinants and Noncolliding Brownian Bridges

Abstract In Chap. 2, for the one-dimensional torus $\mathbb{T}$ and the interval $[0, \pi]$, we showed that the transition probability densities of a single Brownian motion are expressed using the theta function θ. As realizations of polynomials of θ, R_n theta functions were introduced in Chap. 3, where R_n represents the n-th system in one of the seven infinite series of irreducible reduced affine root systems. Rosengren and Schlosser proved that the R_n theta functions give determinantal expressions for the Macdonald denominators. In the present chapter, we give a probability-theoretical interpretation to the determinants of Rosengren and Schlosser; they are proportional to the Karlin–McGregor–Lindström–Gessel–Viennot determinants which provide the total probability masses of n-tuples of noncolliding Brownian paths. Corresponding to the seven types of R_n, we construct seven types of n-particle stochastic processes on $\mathbb{T}$ or in $[0, \pi]$ defined for a finite time duration $[0, T]$. The obtained interacting particle systems are temporally inhomogenous processes called the noncolliding Brownian bridges. The limit $p \to 0$, which causes reduction from the elliptic level to the trigonometric level, corresponds to the temporally homogeneous limit $T \to \infty$. A degeneracy is observed in this limit transition such that the seven processes are reduced to four types of stationary processes of noncolliding Brownian motions.

4.1 Karlin–McGregor–Lindström–Gessel–Viennot (KMLGV) Determinants

We recall the transition probability density $p_{\mathbb{T}}$ of the Brownian motion on $\mathbb{T}$ studied in Sect. 2.2, which has the following expression using the theta function:

$$p_{\mathbb{T}}(0, x; t, y) = \frac{(p_t; p_t)_\infty}{2\pi} \theta(-p_t^{1/2} e^{i(y-x)}; p_t), \quad t > 0, \quad x, y \in [0, 2\pi), \quad (4.1)$$

where

$$p_t := e^{-t}, \quad t > 0. \quad (4.2)$$

© The Author(s), under exclusive license to Springer Nature Singapore Pte Ltd. 2023
M. Katori, *Elliptic Extensions in Statistical and Stochastic Systems*,
SpringerBriefs in Mathematical Physics 47,
https://doi.org/10.1007/978-981-19-9527-9_4

Here we will show an interesting relationship between $p_{\mathbb{T}}$ and the A_{n-1} theta functions $\{\psi_j^{A_{n-1}}\}_{j=1}^n$ studied in Sect. 3.1. At first we assume that

$$n \text{ is an odd integer,}$$

and a set of n points on $\mathbb{T}$ placed with equal distances,

$$\mathbf{u} = \mathbf{u}^{A_{n-1}} = (u_1^{A_{n-1}}, \ldots, u_n^{A_{n-1}})$$

$$\text{with} \quad u_j = u_j^{A_{n-1}} := \frac{2\pi}{n}(j-1), \quad j = 1, \ldots, n. \tag{4.3}$$

We introduce

$$\varpi_{jk}(p_t) = \varpi_{jk}^{A_{n-1}}(p_t) := \frac{1}{(p_{n^2 t}; p_{n^2 t})_\infty} \frac{2\pi}{n} p_t^{-(j-1)^2/2} e^{i(j-1)u_k}, \tag{4.4}$$

$j, k = 1, \ldots, n$. Here notice that $p_{n^2 t} = p_t^{n^2} = e^{-n^2 t}$. Then the following equality holds [16, pp. 216–217], [38].

Lemma 4.1 *Assume that n is odd. Then for $j = 1, \ldots, n$,*

$$\sum_{\ell=1}^n \varpi_{j\ell}(p_t) p_{\mathbb{T}}(0, u_\ell; t, x) = \psi_j^{A_{n-1}}(e^{ix}; p_{nt}, -p_{n^2 t/2}), \quad t > 0, \quad x \in [0, 2\pi).$$

Proof By (4.1) and (4.4), we have

$$\sum_{\ell=1}^n \varpi_{j\ell}(p_t) p_{\mathbb{T}}(0, u_\ell; t, x)$$

$$= \frac{1}{(p_{n^2 t}; p_{n^2 t})_\infty} \frac{1}{n} \sum_{\ell=1}^n p_t^{-(j-1)^2/2} e^{i(j-1)u_\ell} \sum_{m \in \mathbb{Z}} p_t^{\binom{m}{2}} p_t^{m/2} e^{i(x-u_\ell)m}$$

$$= \frac{p_t^{-(j-1)^2/2}}{(p_{n^2 t}; p_{n^2 t})_\infty} \sum_{m \in \mathbb{Z}} p_t^{m^2/2} e^{imx} \frac{1}{n} \sum_{\ell=1}^n e^{i(j-1-m)u_\ell}.$$

Here we use the identity

$$\frac{1}{n} \sum_{\ell=1}^n e^{iju_\ell} = \sum_{k \in \mathbb{Z}} \mathbf{1}(j + nk = 0), \quad j \in \mathbb{Z}. \tag{4.5}$$

Then

$$\sum_{\ell=1}^{n} \varpi_{j\ell}(p_t)\mathrm{p}_{\mathbb{T}}(0, u_\ell; t, x) = \frac{p_t^{-(j-1)^2/2}}{(p_{n^2 t}; p_{n^2 t})_\infty} \sum_{m\in\mathbb{Z}} p_t^{m^2/2} e^{imx} \sum_{k\in\mathbb{Z}} \mathbf{1}(m = j - 1 + kn)$$

$$= \frac{p_t^{-(j-1)^2/2}}{(p_{n^2 t}; p_{n^2 t})_\infty} \sum_{k\in\mathbb{Z}} p_t^{(j-1+kn)^2/2} e^{i(j-1+kn)x}.$$

We see that

$$p_t^{(j-1+kn)^2/2} = p_t^{(j-1)^2/2+n^2 k(k-1)/2+n(j-1)k+n^2 k/2} = p_t^{(j-1)^2/2} p_{nt}^{n\binom{k}{2}} (p_{nt}^{j-1} p_{n^2 t/2})^k.$$

Hence we have

$$\sum_{\ell=1}^{n} \varpi_{j\ell}(p_t)\mathrm{p}_{\mathbb{T}}(0, u_\ell; t, x) = \frac{e^{i(j-1)x}}{(p_{n^2 t}; p_{n^2 t})_\infty} \sum_{k\in\mathbb{Z}} (-1)^k p_{nt}^{n\binom{k}{2}} (-p_{nt}^{j-1} p_{n^2 t/2} e^{inx})^k.$$

Since $(-1)^{n-1} = 1$ when n is odd, the above is equal to

$$e^{i(j-1)x} \theta((-1)^{n-1} p_{nt}^{j-1}(-p_{n^2 t/2}) e^{inx}; p_{nt}^n),$$

and hence the assertion is concluded by the definition (3.1) of $\psi_j^{A_{n-1}}(\zeta; p, r)$.   □

Now we introduce a configuration space of n points on $\mathbb{T}$ in which the locations of n points are distinct and ordered as follows:

$$\mathbb{W}_n([0, 2\pi)) := \{\mathbf{x} = (x_1, \ldots, x_n) \in \mathbb{R}^n : 0 \le x_1 < \cdots < x_n < 2\pi\}.$$

This is called the *Weyl alcove* of type A_{n-1}. The equidistance points $\mathbf{u}$ given by (4.3) is in $\mathbb{W}_n([0, 2\pi))$. Assume that $\mathbf{x} \in \mathbb{W}_n([0, 2\pi))$. Then Lemma 4.1 provides the following determinantal equality:

$$\det_{1\le j,\ell\le n}[\varpi_{j\ell}(p_t)] \det_{1\le\ell,m\le n}[\mathrm{p}_{\mathbb{T}}(0, u_\ell; t, x_m)] = \det_{1\le j,m\le n}[\psi_j^{A_{n-1}}(e^{ix_m}; p_{nt}, -p_{n^2 t/2})].$$

We can calculate as (see [16, p. 216])

$$\det_{1\le j,\ell\le n}[\varpi_{j\ell}(p_t)] = (-i)^{(n-1)/2} \frac{(2\pi)^n}{n^{n/2}} \frac{p_t^{-n(n-1)(2n-1)/12}}{(p_{n^2 t}; p_{n^2 t})_\infty^n}$$

$$=: (-i)^{(n-1)/2} \varpi^{A_{n-1}}(p_t). \tag{4.6}$$

Hence we have the following formula for the determinant consisting of the basis of the A_{n-1} theta functions using the transition probability densities of the Brownian motions on $\mathbb{T}$:

$$\det_{1 \le j,k \le n} [\psi_j^{A_{n-1}}(e^{ix_k}; p_{nt}, -p_{n^2t/2})]$$

$$= (-i)^{(n-1)/2} \varpi^{A_{n-1}}(p_t) \det_{1 \le j,k \le n} [p_{\mathbb{T}}(0, u_j; t, x_k)],$$

$$t > 0, \quad \mathbf{x} \in \mathbb{W}_n([0, 2\pi)), \tag{4.7}$$

where n is odd. The determinant of transition probability densities is known as the *Karlin–McGregor–Lindström–Gessel–Viennot determinant* and has the following probabilistic meaning [22, 32, 53] (see, for instance, Sect. 3.3 in [36]); provided $\mathbf{u}, \mathbf{x} \in \mathbb{W}_n([0, 2\pi))$:

if n is odd,

$$\det_{1 \le j,k \le n} [p_{\mathbb{T}}(0, u_j; t, x_k)] = \begin{bmatrix} \text{the total probability mass of } n\text{-tuples} \\ \text{of noncolliding Brownian paths on} \\ \mathbb{T} \times [0, t] \text{ from } \mathbf{u} \text{ at time } 0 \text{ to } \mathbf{x} \text{ at time } t \end{bmatrix} > 0.$$
$$\tag{4.8}$$

It should be noticed that such a probabilistic interpretation holds if and only if n is odd, when we consider the noncolliding paths on $\mathbb{T}$. See Fig. 4.1.

If n is even, we have to replace the transition probability density $p_{\mathbb{T}}$ by $\widetilde{p}_{\mathbb{T}}$. It was defined by (2.28) as an alternative sum of the transition probability densities of the Brownian motion on $\mathbb{R}$, in which the sign changes depending on the parity

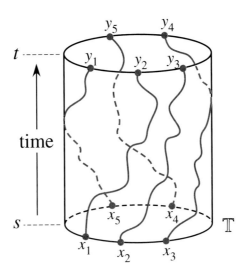

Fig. 4.1 Schematic picture of 5-tuples of noncolliding Brownian paths on the spatio-temporal surface $\mathbb{T} \times [s, t]$. The configuration $\mathbf{x}$ at time s and $\mathbf{y}$ at t are both in $\mathbb{W}_n([0, 2\pi))$. Since the paths can be wound around the one-dimensional torus (i.e., circle) $\mathbb{T}$, the indices of $\mathbf{y}$ are shifted from those of $\mathbf{x}$ with modulus n. In this figure, for $j = 1, \ldots, 4$ the end point at time t of the j-th path starting from x_j at time s is denoted by y_{j+1}, and that of the fifth path from x_5 is given by y_1

of the winding number of the path on the torus $\mathbb{T}$ [20, 52, 56]. That is, provided $\mathbf{u}, \mathbf{x} \in \mathbb{W}_n([0, 2\pi))$,

if n is even,

$$\det_{1 \le j,k \le n} [\widetilde{p}_\mathbb{T}(0, u_j; t, x_k)] = \begin{bmatrix} \text{the total probability mass of } n\text{-tuples} \\ \text{of noncolliding Brownian paths on} \\ \mathbb{T} \times [0, t] \text{ from } \mathbf{u} \text{ at time } 0 \text{ to } \mathbf{x} \text{ at time } t \end{bmatrix} > 0.$$

(4.9)

For even n, $\{\varpi_{jk}(p_t)\}_{1 \le j,k \le n}$ given by (4.4) are replaced by

$$\widetilde{\varpi}_{jk}(p_t) = \widetilde{\varpi}_{jk}^{A_{n-1}}(p_t) := \frac{1}{(p_{n^2t}; p_{n^2t})_\infty} \frac{2\pi}{n} p_t^{-(j-1/2)^2/2} e^{i(j-1/2)u_k},$$

(4.10)

$j, k = 1, \ldots, n$, which gives [16, p. 216]

$$\det_{1 \le j,k \le n} [\widetilde{\varpi}_{jk}(p_t)] = (-i)^{n/2} \frac{(2\pi)^n}{n^{n/2}} \frac{p_t^{-n(2n-1)(2n+1)/24}}{(p_{n^2t}; p_{n^2t})_\infty^n}$$
$$=: (-i)^{n/2} \widetilde{\varpi}^{A_{n-1}}(p_t).$$

(4.11)

Instead of Lemma 4.1, the following equality is proved [16, pp. 216–217], [38].

Lemma 4.2 *Assume that n is even. Then for $j = 1, \ldots, n$, $t > 0$, $x \in [0, 2\pi]$,*

$$\sum_{\ell=1}^n \widetilde{\varpi}_{j\ell}(p_t)\widetilde{p}_\mathbb{T}(0, u_\ell; t, x) = e^{ix/2}\psi_j^{A_{n-1}}(e^{ix}; p_{nt}, p_{n(n+1)t/2}).$$

Proof By (2.29) and (4.10), we have

$$\sum_{\ell=1}^n \widetilde{\varpi}_{j\ell}(p_t)\widetilde{p}_\mathbb{T}(0, u_\ell; t, x) = \frac{p_t^{-j(j-1)/2} e^{-ix/2}}{(p_{n^2t}; p_{n^2t})_\infty} \sum_{m \in \mathbb{Z}} p_t^{m(m-1)/2} e^{imx} \frac{1}{n} \sum_{\ell=1}^n e^{i(j-m)u_\ell}.$$

Using (4.5), the above is written as

$$\frac{p_t^{-j(j-1)/2} e^{-ix/2}}{(p_{n^2t}; p_{n^2t})_\infty} \sum_{k \in \mathbb{Z}} p_t^{(j+kn)(j+kn-1)/2} e^{i(j+kn)x}.$$

We see that

$$p_t^{(j+kn)(j+kn-1)/2} = p_t^{j(j-1)/2} p_{nt}^{n\binom{k}{2}} (p_{nt}^{j-1} p_{n(n+1)t/2})^k.$$

Hence we have

$$\sum_{\ell=1}^{n} \widetilde{\varpi}_{j\ell}(p_t)\widetilde{p}_{\mathbb{T}}(0, u_\ell; t, x) = \frac{e^{ix/2}e^{i(j-1)x}}{(p_{n^2 t}; p_{n^2 t})_\infty} \sum_{k\in\mathbb{Z}}(-1)^k p_{nt}^{n\binom{k}{2}}(-p_{nt}^{j-1} p_{n(n+1)t/2}e^{inx})^k$$

$$= e^{ix/2}e^{i(j-1)x}\theta(-p_{nt}^{j-1} p_{n(n+1)t/2}e^{inx}; p_{nt}^n).$$

Since $(-1)^{n-1} = -1$ when n is even, this proves the assertion by the definition (3.1). $\square$

The following equality is thus proved, when n is even; for $t > 0$, $\mathbf{x} \in \mathbb{W}_n([0, 2\pi))$:

$$\det_{1\leq j,k\leq n} [\psi_j^{A_{n-1}}(e^{ix_k}; p_{nt}, p_{n(n+1)t/2})]$$

$$= (-i)^{n/2}e^{-i\sum_{\ell=1}^{n} x_\ell/2}\widetilde{\varpi}^{A_{n-1}}(p_t) \det_{1\leq j,k\leq n} [\widetilde{p}_{\mathbb{T}}(0, u_j; t, x_k)]. \tag{4.12}$$

4.2   Noncolliding Brownian Bridges on $\mathbb{T}$

Let $T > 0$. Then for $t \in [0, T]$, $\mathbf{x} \in \mathbb{W}_n([0, 2\pi))$, define

$$\mathbf{p}_T^{A_{n-1}}(t, \mathbf{x}) = \mathbf{p}_T^{A_{n-1}}(0, \mathbf{u}; t, \mathbf{x}; T, \mathbf{u})$$

$$:= \begin{cases} \dfrac{\det\limits_{1\leq j,k\leq n} [p_{\mathbb{T}}(0, u_j; t, x_k)] \det\limits_{1\leq j,k\leq n} [p_{\mathbb{T}}(t, x_j; T, u_k)]}{\det\limits_{1\leq j,k\leq n} [p_{\mathbb{T}}(0, u_j; T, u_k)]}, \\ \qquad\qquad\qquad\qquad\qquad\qquad\qquad\qquad \text{if } n \text{ is odd,} \\[2em] \dfrac{\det\limits_{1\leq j,k\leq n} [\widetilde{p}_{\mathbb{T}}(0, u_j; t, x_k)] \det\limits_{1\leq j,k\leq n} [\widetilde{p}_{\mathbb{T}}(t, x_j; T, u_k)]}{\det\limits_{1\leq j,k\leq n} [\widetilde{p}_{\mathbb{T}}(0, u_j; T, u_k)]}, \\ \qquad\qquad\qquad\qquad\qquad\qquad\qquad\qquad \text{if } n \text{ is even,} \end{cases} \tag{4.13}$$

where $\mathbf{u} = \mathbf{u}^{A_{n-1}}$ given by (4.3). The probability theoretical interpretations of the first factors in the numerators were already given by (4.8) and (4.9). Similarly, the second factors in the numerators give the total probability mass of n-tuples of noncolliding Brownian paths on $\mathbb{T}$ from $\mathbf{x}$ at time t to $\mathbf{u}$ at time T. The products of them are divided by the total probability masses of the paths on $\mathbb{T}$ from $\mathbf{u}$ at time 0 returning to $\mathbf{u}$ at time T. Hence the positivity is guaranteed,

$$\mathbf{p}_T^{A_{n-1}}(t, \mathbf{x}) > 0, \quad t \in [0, T], \quad \mathbf{x} \in \mathbb{W}_n([0, 2\pi)). \tag{4.14}$$

Moreover, $\mathbf{p}_T^{A_{n-1}}$ is well normalized as mentioned by Proposition 4.1 below.

In order to prove Proposition 4.1, first we show the equality known as the *Andréief identity* [17]. Here we consider the following general setting. Let a space S be a subset of $\mathbb{R}^d$ with $d \in \mathbb{N}$ equipped with a *reference measure* λ on it. Examples of S are the real line $\mathbb{R}$ with the Lebesgue measure $\lambda(dx) = dx$, the complex plane $\mathbb{C} \simeq \mathbb{R}^2$ with $\lambda(dz) = dxdy$ for $z = x + iy \in \mathbb{C}$, the unit circle $\mathbb{T} = \{e^{ix} : x \in [0, 2\pi)\}$ with the uniform measure $\lambda_{\mathbb{T}}(dx) := dx/2\pi$, and an interval $[0, \pi] \subset \mathbb{R}$ with $\lambda_{[0,\pi]}(dx) := dx/\pi$, and so on.

Lemma 4.3 *Fix $n \in \mathbb{N}$. Assume that $\{f_j(x)\}_{j=1}^n$ and $\{g_j(x)\}_{j=1}^n$ are two sets of square integrable functions of $x \in S$, each of which consists of linearly independent functions. Then the following equality holds:*

$$\frac{1}{n!} \int_{S^n} \det_{1 \le j,k \le n} [f_j(x_k)] \det_{1 \le j,k \le n} [g_j(x_k)] \lambda(d\mathbf{x})$$

$$= \det_{1 \le j,k \le n} \left[\int_S f_j(x) g_k(x) \lambda(dx) \right], \tag{4.15}$$

where $\lambda(d\mathbf{x})$ is the n-direct product of reference measures; $\lambda(d\mathbf{x}) := \prod_{j=1}^n \lambda(dx_j)$.

Proof By the definition of a *determinant*, if we write the set of all permutations of a series of indices $(1, \ldots, n)$ as $\mathfrak{S}_n$, we have

$$\det_{1 \le j,k \le n} [f_j(x_k)] = \sum_{\sigma \in \mathfrak{S}_n} \mathrm{sgn}(\sigma) \prod_{j=1}^n f_j(x_{\sigma(j)}),$$

$$\det_{1 \le j,k \le n} [g_j(x_k)] = \sum_{\rho \in \mathfrak{S}_n} \mathrm{sgn}(\rho) \prod_{k=1}^n g_{\rho(k)}(x_k),$$

where $\mathrm{sgn}(\sigma)$ denotes the signature of permutation σ. Since for each $\sigma, \rho \in \mathfrak{S}_n$,

$$\prod_{j=1}^n f_j(x_{\sigma(j)}) \prod_{k=1}^n g_{\rho(k)}(x_k) = \prod_{j=1}^n f_j(x_{\sigma(j)}) g_{\rho(\sigma(j))}(x_{\sigma(j)}),$$

and $\mathrm{sgn}(\sigma)\mathrm{sgn}(\rho) = \mathrm{sgn}(\rho \circ \sigma)$, if we write $\rho \circ \sigma =: \kappa \in \mathfrak{S}_n$, then the following equality is concluded:

$$\det_{1 \le j,k \le n} [f_j(x_k)] \det_{1 \le j,k \le n} [g_j(x_k)] = \sum_{\sigma \in \mathfrak{S}_n} \sum_{\kappa \in \mathfrak{S}_n} \mathrm{sgn}(\kappa) \prod_{j=1}^n f_j(x_{\sigma(j)}) g_{\kappa(j)}(x_{\sigma(j)}).$$

We integrate both sides of this equality with respect to the reference measure over S^n and divide them by $n!$. Then the LHS gives the integral

$$\frac{1}{n!} \int_{S^n} \det_{1 \le j,k \le n} [f_j(x_k)] \det_{1 \le j,k \le n} [g_j(x_k)] \lambda(d\mathbf{x})$$

and the RHS gives

$$\sum_{\kappa \in \mathfrak{S}_n} \operatorname{sgn}(\kappa) \sum_{\sigma \in \mathfrak{S}_n} \frac{1}{n!} \int_{S^n} \prod_{j=1}^n f_j(x_{\sigma(j)}) g_{\kappa(j)}(x_{\sigma(j)}) \lambda(d\mathbf{x}).$$

We see that for each $\kappa \in \mathfrak{S}_n$,

$$\sum_{\sigma \in \mathfrak{S}_n} \frac{1}{n!} \int_{S^n} \prod_{j=1}^n f_j(x_{\sigma(j)}) g_{\kappa(j)}(x_{\sigma(j)}) \lambda(d\mathbf{x})$$

$$= \int_{S^n} \prod_{j=1}^n f_j(x_j) g_{\kappa(j)}(x_j) \lambda(d\mathbf{x}) = \prod_{j=1}^n \int_S f_j(x) g_{\kappa(j)}(x) \lambda(dx).$$

Therefore, we obtain the equality (4.15), and the proof is complete. $\square$

Proposition 4.1 *For each $t \in [0, T]$,*

$$\int_{\mathbb{W}_n([0,2\pi))} \mathbf{p}_T^{A_{n-1}}(t, \mathbf{x}) d\mathbf{x} = 1, \tag{4.16}$$

where $d\mathbf{x}$ denotes the Lebesgue measure; $d\mathbf{x} := \prod_{j=1}^n dx_j$. Hence, $\mathbf{p}_T^{A_{n-1}}(t, \cdot)$ gives a probability density of n point configuration $\mathbf{x} \in \mathbb{W}_n([0, 2\pi))$ at each $t \in [0, T]$.

Proof When $S = \mathbb{T}$, the Andréief identity (4.15) gives

$$\int_{\mathbb{W}_n([0,2\pi))} \det_{1 \leq j,k \leq n} [f_j(x_k)] \det_{1 \leq j,k \leq n} [g_j(x_k)] d\mathbf{x} = \det_{1 \leq j,k \leq n} \left[\int_{\mathbb{T}} f_j(x) g_k(x) dx \right]. \tag{4.17}$$

Now we assume that n is odd. If we apply (4.17) to the integral of the numerator in the RHS of (4.13), we have

$$\int_{\mathbb{W}_n([0,2\pi))} \det_{1 \leq j,k \leq n} [p_{\mathbb{T}}(0, u_j; t, x_k)] \det_{1 \leq j,k \leq n} [p_{\mathbb{T}}(t, x_j; T, u_k)] d\mathbf{x}$$

$$= \det_{1 \leq j,k \leq n} \left[\int_{\mathbb{T}} p_{\mathbb{T}}(0, u_j; t, x) p_{\mathbb{T}}(t, x; T, u_k) dx \right].$$

We use the Chapman–Kolmogorov equation (2.14) and find that this is exactly equal to the denominator $\det_{1 \leq j,k \leq n} [p_{\mathbb{T}}(0, u_j; T, u_k)]$. Hence the assertion (4.16) is proved. We can prove this also in the case with even n by the Chapman–Kolmogorov equation (2.31) for $\widetilde{p}_{\mathbb{T}}$. Combining with the positivity (4.14), we can conclude that $\mathbf{p}_T^{A_{n-1}}(t, \cdot)$ gives a probability density of n point configuration $\mathbf{x} \in \mathbb{W}_n([0, 2\pi))$ at each $t \in [0, T]$. $\square$

We consider T as a given *time duration* and $\mathbf{p}_T^{A_{n-1}}(t, \mathbf{x})$ given by (4.13) determines the probability distribution of an n-particle configuration $\mathbf{x} = (x_1, \ldots, x_n)$ in $\mathbb{T}$ at time $t \in [0, T]$ of a *Markov process*. We assume that $\mathbf{p}^{A_{n-1}}(t, \mathbf{x}) \equiv 0$ if $\mathbf{x} \notin \mathbb{W}_n([0, 2\pi))$ at any $t \in [0, T]$, and hence the process consists of n-tuples of noncolliding Brownian paths on $\mathbb{T}$. By (4.13), both of the initial configuration at time $t = 0$ and the final configuration at time $t = T$ are $\mathbf{u} \in \mathbb{W}_n([0, 2\pi))$ given by (4.3). This process is called the n-particle system of *noncolliding Brownian bridges*. A more precise definition is given as follows.[1]

Definition 4.1 Fix $T > 0$ and the initial and the final configurations are identical and given by $\mathbf{u}$ defined by (4.3). The system of *noncolliding Brownian bridges* of type A_{n-1} is the Markov process $(\mathbf{X}^{A_{n-1}}(t))_{t \in [0,T]}$ of n particles on $\mathbb{T}$ with time duration T, such that its is determined by the following density function: For an arbitrary $m \in \mathbb{N}$ and for an arbitrary series of strictly increasing times $t_0 := 0 < t_1 < \cdots < t_m < T$, provided that $\mathbf{x}^{(0)} := \mathbf{u}$, the *multi-time joint probability density* of configurations $\mathbf{x}^{(\ell)} \in \mathbb{W}_n([0, 2\pi))$ at times $t = t_\ell, \ell = 1, \ldots, m$ is given by

$$\mathbf{p}_T^{A_{n-1}}(t_1, \mathbf{x}^{(1)}; t_2, \mathbf{x}^{(2)}; \cdots; t_m, \mathbf{x}^{(m)})$$

$$= \prod_{\ell=1}^{m} \det_{1 \leq j,k \leq n} [\mathrm{p}_{\mathbb{T}}(t_{\ell-1}, x_j^{(\ell-1)}; t_\ell, x_k^{(\ell)})] \frac{\det_{1 \leq j,k \leq n} [\mathrm{p}_{\mathbb{T}}(t_m, x_j^{(m)}; T, u_k)]}{\det_{1 \leq j,k \leq n} [\mathrm{p}_{\mathbb{T}}(0, u_j; T, u_k)]}, \quad (4.18)$$

if n is odd. If n is even, the multi-time joint probability density is given by the formula (4.18), replacing $\mathrm{p}_{\mathbb{T}}$ by $\widetilde{\mathrm{p}}_{\mathbb{T}}$.

By Proposition 4.1, the following *reducibility of joint probability densities* is readily verified:

$$\int_{\mathbb{W}_n([0,2\pi))} \mathbf{p}_T^{A_{n-1}}(t_1, \mathbf{x}^{(1)}; \cdots; t_{\ell-1}, \mathbf{x}^{(\ell-1)}; t_\ell, \mathbf{x}^{(\ell)}; t_{\ell+1}, \mathbf{x}^{(\ell+1)}; \cdots; t_m, \mathbf{x}^{(m)}) d\mathbf{x}^{(\ell)}$$

$$= \mathbf{p}_T^{A_{n-1}}(t_1, \mathbf{x}^{(1)}; \cdots; t_{\ell-1}, \mathbf{x}^{(\ell-1)}; t_{\ell+1}, \mathbf{x}^{(\ell+1)}; \cdots; t_m, \mathbf{x}^{(m)}), \quad \ell = 1, \ldots, m.$$

By the biorthogonality of the A_{n-1} theta functions given by Proposition 3.3 and the equalities (4.7) with (4.6) and (4.12) with (4.11), we can prove the following. Let

$$r_t = r_t(n) := \begin{cases} -p n^2 t/2, & \text{if } n \text{ is odd}, \\ p n(n+1)t/2, & \text{if } n \text{ is even}. \end{cases} \quad (4.19)$$

---

[1] The construction and the basic properties of the *Brownian bridge* on $\mathbb{R}$ are summarized in [3, IV.4]. Its n-particle extension is defined by Definition 4.1 by specifying the *finite-dimensional distributions* (see [3, IV.4.23]). See also Sect. 7.1 of the present monograph for stochastic differential equations of Brownian bridges.

Proposition 4.2 *Let*

$$
Z^{A_{n-1}} = Z^{A_{n-1}}(p_t, p_T) := (2\pi)^n \prod_{\ell=1}^{n} h_\ell^{A_{n-1}}(p_{nt}, p_{n(T-t)}, |r_T|)
$$

$$
= \frac{(2\pi)^n \theta(-|r_T|; p_{nT})(p_{n^2 T}; p_{n^2 T})_\infty^n}{(p_{n^2 t}; p_{n^2 t})_\infty^n (p_{n^2(T-t)}; p_{n^2(T-t)})_\infty^n}, \quad t \in [0, T].
$$

(4.20)

Then for each $t \in [0, T]$,

$$
\mathbf{p}_T^{A_{n-1}}(t, \mathbf{x})
$$

$$
= \frac{1}{Z^{A_{n-1}}} \det_{1\le j,k\le n} [\psi_j^{A_{n-1}}(e^{ix_k}; p_{nt}, r_t)] \, \overline{\det_{1\le j,k\le n} [\psi_j^{A_{n-1}}(e^{ix_k}; p_{n(T-t)}, r_{T-t})]}
$$

$$
= \frac{1}{Z^{A_{n-1}}} \overline{\det_{1\le j,k\le n} [\psi_j^{A_{n-1}}(e^{ix_k}; p_{nt}, r_t)]} \, \det_{1\le j,k\le n} [\psi_j^{A_{n-1}}(e^{ix_k}; p_{n(T-t)}, r_{T-t})],
$$

$$
\mathbf{x} \in \mathbb{W}_n([0, 2\pi)).
$$

(4.21)

Proof First we consider the case with odd n. The last equality in (4.7) with (4.6) gives

$$
\det_{1\le j,k\le n} [p_T(0, u_j; t, x_k)] = \frac{i^{(n-1)/2}}{\varpi^{A_{n-1}}(p_t)} \det_{1\le j,k\le n} [\psi_j^{A_{n-1}}(e^{ix_k}; p_{nt}, -p_{n^2 t/2})].
$$

We replace t by $T - t$ in the above equation. By (2.13), $p_T(0, u_j; T - t, x_k) = p_T(t, u_j; T, x_k) = p_T(t, x_k; T, u_j)$, and hence we have

$$
\det_{1\le j,k\le n} [p_T(t, x_j; T, u_k)]
$$

$$
= \frac{i^{(n-1)/2}}{\varpi^{A_{n-1}}(p_{T-t})} \det_{1\le j,k\le n} [\psi_j^{A_{n-1}}(e^{ix_k}; p_{n(T-t)}, -p_{n^2(T-t)/2})].
$$

Since the KMLGV determinant in the LHS is positive and (4.6) defines $\varpi^{A_{n-1}}(p_{T-t})$ as a real-valued function, the complex conjugate of the above is written

$$
\det_{1\le j,k\le n} [p_T(t, x_j; T, u_k)]
$$

$$
= \frac{(-i)^{(n-1)/2}}{\varpi^{A_{n-1}}(p_{T-t})} \det_{1\le j,k\le n} \overline{[\psi_j^{A_{n-1}}(e^{ix_k}; p_{n(T-t)}, -p_{n^2(T-t)/2})]}.
$$

Hence the definition (4.13) gives

$$
\mathbf{p}_T^{A_{n-1}}(t, \mathbf{x}) = \frac{1}{Z} \det_{1 \le j,k \le n} [\psi_j^{A_{n-1}}(e^{ix_k}; p_{nt}, -p_{n^2 t}/2)]
$$

$$
\times \det_{1 \le j,k \le n} \overline{[\psi_j^{A_{n-1}}(e^{ix_k}; p_{n(T-t)}, -p_{n^2(T-t)}/2)]}, \tag{4.22}
$$

where $Z = \varpi^{A_{n-1}}(p_t)\varpi^{A_{n-1}}(p_{T-t}) \det_{1 \le j,k \le n} [\mathrm{p}_{\mathbb{T}}(0, u_j; T, u_k)] \in \mathbb{R}$. If we calculate the integrals of both sides of the equality (4.22) with respect to $\mathbf{x}$ over $\mathbb{W}_n([0, 2\pi))$, the LHS gives 1 by Proposition 4.1. Applying the Andréief identity (4.17), the RHS gives

$$
\frac{(2\pi)^n}{Z} \det_{1 \le j,k \le n} \left[\int_0^{2\pi} \frac{dx}{2\pi} \psi_j^{A_{n-1}}(e^{ix}; p_{nt}, -p_{n^2 t}/2) \overline{\psi_k^{A_{n-1}}(e^{ix}; p_{n(T-t)}, -p_{n^2(T-t)}/2)} \right]
$$

$$
= \frac{(2\pi)^n}{Z} \det_{1 \le j,k \le n} \left[h_j^{A_{n-1}}(p_{nt}, p_{n(T-t)}, p_{n^2 T}/2) \delta_{jk} \right]
$$

$$
= \frac{(2\pi)^n}{Z} \prod_{j=1}^n h_j^{A_{n-1}}(p_{nt}, p_{n(T-t)}, p_{n^2 T}/2).
$$

Hence $Z = (2\pi)^n \prod_{j=1}^n h_j^{A_{n-1}}(p_{nt}, p_{n(T-t)}, p_{n^2 T}/2)$.

Next we consider the case with even n. The last equality in (4.12) gives

$$
\det_{1 \le j,k \le n} [\widetilde{\mathrm{p}}_{\mathbb{T}}(0, u_j; t, x_k)] = \frac{i^{n/2} e^{i\sum_{\ell=1}^n x_\ell}}{\widetilde{\varpi}^{A_{n-1}}(p_t)} \det_{1 \le j,k \le n} [\psi_j^{A_{n-1}}(e^{ix_k}; p_{nt}, p_{n(n+1)t}/2)].
$$

By the same argument as given above, we have

$$
\det_{1 \le j,k \le n} [\widetilde{\mathrm{p}}_{\mathbb{T}}(t, x_j; T, u_k)]
$$

$$
= \frac{(-i)^{n/2} e^{-i\sum_{\ell=1}^n x_\ell}}{\widetilde{\varpi}^{A_{n-1}}(p_{T-t})} \det_{1 \le j,k \le n} \overline{[\psi_j^{A_{n-1}}(e^{ix_k}; p_{n(T-t)}, p_{n(n+1)(T-t)}/2)]}.
$$

Hence the definition (4.13) gives

$$
\mathbf{p}_T^{A_{n-1}}(t, \mathbf{x}) = \frac{1}{Z} \det_{1 \le j,k \le n} [\psi_j^{A_{n-1}}(e^{ix_k}; p_{nt}, p_{n(n+1)t}/2)]
$$

$$
\times \det_{1 \le j,k \le n} \overline{[\psi_j^{A_{n-1}}(e^{ix_k}; p_{n(T-t)}, p_{n(n+1)(T-t)}/2)]}, \tag{4.23}
$$

where $Z = \widetilde{\varpi}^{A_{n-1}}(p_t)\widetilde{\varpi}^{A_{n-1}}(p_{T-t})\det_{1\le j,k\le n}[\widetilde{p}_{\mathbb{T}}(0, u_j; T, u_k)] \in \mathbb{R}$. As we did for the

case with odd n, we can determine $Z = (2\pi)^n \prod_{j=1}^{n} h_j^{A_{n-1}}(p_{nt}, p_{n(T-t)}, p_{n(n+1)T/2})$. By

(3.24) in Proposition 3.3, we have

$$Z = \frac{(2\pi)^n (p_{n^2T}; p_{n^2T})_\infty^n}{(p_{n^2t}; p_{n^2t})_\infty^n (p_{n^2(T-t)}; p_{n^2(T-t)})_\infty^n} \prod_{j=1}^{n} \theta(-|r_T|p_{nT}^{j-1}; p_{nT}^n)$$

$$= \frac{(2\pi)^n (p_{n^2T}; p_{n^2T})_\infty^n}{(p_{n^2t}; p_{n^2t})_\infty^n (p_{n^2(T-t)}; p_{n^2(T-t)})_\infty^n} \theta(-|r_T|; p_{nT}),$$

where we have used (1.12). Hence the proof of (4.21) is complete.   □

4.3 KMLGV Determinants and Noncolliding Brownian Bridges in $[0, \pi]$

For the spaces of other R_n theta functions, $\mathcal{E}_{p_t}^{R_n}$ with $p_t = e^{-t}, t > 0$, we consider the interval $[0, \pi]$ representing the half perimeter of the circle $\mathbb{T}$ and the Weyl alcove,

$$\mathbb{W}_n([0, \pi]) := \{\mathbf{x} \in \mathbb{R}^n : 0 \le x_1 < \cdots < x_n \le \pi\}.$$

We choose the following equidistance configurations of n particles in $\mathbb{W}_n([0, \pi])$:

$$\mathbf{u} = \mathbf{u}^{R_n} = (u_1^{R_n}, \ldots, u_n^{R_n})$$

$$\text{with }\ u_j^{R_n} := \frac{2\pi}{\mathcal{N}} \times \begin{cases} j - 1/2, & \text{for } R_n = B_n, B_n^\vee, \\ j, & \text{for } R_n = C_n, C_n^\vee, BC_n, \\ j - 1, & \text{for } R_n = D_n, \end{cases} \qquad (4.24)$$

where $\mathcal{N} = \mathcal{N}^{R_n}$ given by (3.9). See Fig. 4.2. Then we define $\{\varpi_{jk}^{R_n}(p_t)\}_{1\le j,k\le n}$ as follows. For $j, k = 1, \ldots, n$,

$$\varpi_{jk}(p_t) = \varpi_{jk}^{R_n}(p_t) := \frac{1}{(p_{\mathcal{N}^2t}; p_{\mathcal{N}^2t})_\infty} \frac{4\pi}{\mathcal{N}} p_t^{-\gamma_j^2/2} \widetilde{c}_k$$

$$\times \begin{cases} i\sin(\gamma_j u_k), & \text{for } R_n = B_n, B_n^\vee, C_n, C_n^\vee, BC_n, \\ \cos(\gamma_j u_k), & \text{for } R_n = D_n, \end{cases}$$

$$(4.25)$$

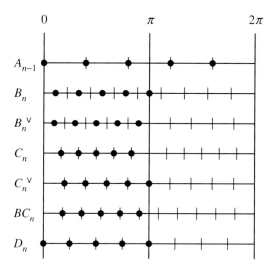

Fig. 4.2 The equidistance configurations $\mathbf{u}^{R_n}$ are shown when $n = 5$. They are used for the initial and final configurations of the noncolliding Brownian bridges. The boundary conditions are summarized by (4.35). Since the boundary condition is (aa) for $B_n^\vee$ and C_n, there should be no point both at 0 and π, while for D_n we can put points both at 0 and π, since the boundary condition is (rr). For $B_n, C_n^\vee, BC_n$, the origin should be vacant, since the boundary condition (a) is imposed at 0

where

$$\gamma_j = \gamma_j^{R_n} := \begin{cases} j - n - 1/2, & R_n = B_n, C_n^\vee, BC_n, \\ j - n - 1, & R_n = B_n^\vee, C_n, \\ j - n, & R_n = D_n, \end{cases}$$

$$\tilde{c}_k = \tilde{c}_k^{R_n} := \begin{cases} 1, & k = 1, \dots, n, & R_n = B_n^\vee, C_n, BC_n, \\ \begin{cases} 1, & k = 1, \dots, n-1, \\ 1/2, & k = n, \end{cases} & R_n = B_n, C_n^\vee, \\ \begin{cases} 1, & k = 2, \dots, n-1, \\ 1/2, & k = 1, n, \end{cases} & R_n = D_n, \end{cases}$$

$\mathcal{N} = \mathcal{N}^{R_n}$ given by (3.9), and $\mathbf{u} = \mathbf{u}^{R_n}$ given by (4.24).

We can prove the following [38].

Lemma 4.4 *For* $j = 1, \ldots, n$, $t > 0$, *and* $x \in [0, \pi]$,

$$\sum_{\ell=1}^{n} \varpi_{j\ell}(p_t) \mathrm{p}_{[0,\pi]}^{\mathrm{ar}}(0, u_\ell; t, x) = e^{-ix/2} \psi_j^{R_n}(e^{ix}; p_{Nt}), \quad \text{if } R_n = B_n, C_n^\vee, BC_n,$$

$$\sum_{\ell=1}^{n} \varpi_{j\ell}(p_t) \mathrm{p}_{[0,\pi]}^{\mathrm{aa}}(0, u_\ell; t, x) = \psi_j^{R_n}(e^{ix}; p_{Nt}), \quad \text{if } R_n = B_n^\vee, C_n,$$

$$\sum_{\ell=1}^{n} \varpi_{j\ell}(p_t) \mathrm{p}_{[0,\pi]}^{\mathrm{rr}}(0, u_\ell; t, x) = \psi_j^{D_n}(e^{ix}; p_{(2n-2)t}), \quad \text{if } R_n = D_n, \qquad (4.26)$$

where $\varpi_{j\ell}(p_t) = \varpi_{j\ell}^{R_n}(p_t)$ *given by (4.25) and* $\mathbf{u} = \mathbf{u}^{R_n}$ *given by (4.24).*

Proof (i) First we prove (4.26) for $R_n = C_n$ and $B_n^\vee$. By (2.20), (4.24), and (4.25), in these cases we have

$$L_j^{R_n} := \sum_{\ell=1}^{n} \varpi_{j\ell}^{R_n}(p_t) \mathrm{p}_{[0,\pi]}^{\mathrm{aa}}(0, u_\ell; t, x)$$

$$= \frac{2i(p_t; p_t)_\infty}{(p_{N^2 t}; p_{N^2 t})_\infty} \frac{p_t^{-\gamma_j^2/2}}{N}$$

$$\times \left\{ \sum_{\ell=1}^{n} \sin(\gamma_j u_\ell) \theta(-p_t^{1/2} e^{i(x-u_\ell)}; p_t) - \sum_{\ell=1}^{n} \sin(\gamma_j u_\ell) \theta(-p_t^{1/2} e^{i(x+u_\ell)}; p_t) \right\}. \tag{4.27}$$

Consider the case that $R_n = C_n$. Since $\sin(\gamma_j^{C_n} u_\ell^{C_n}) = \sin(\gamma_j^{C_n} \pi \ell/(n+1)) = 0$ if $\ell = 0$ and $\ell = n+1$ for $\gamma_j^{C_n} = j - n - 1 \in \mathbb{Z}$, we can change the range of ℓ in the first summation in the bracket from $\{1, \ldots, n\}$ to $\{0, 1, \ldots, n, n+1\}$. In the second summation there, we replace the variable ℓ by $-\ell$ and change the range from $\{1, \ldots, n\}$ to $\{-n, -n+1, \ldots, -1\}$. Then the above is written as

$$L_j^{C_n} = \frac{2i(p_t; p_t)_\infty}{(p_{N^2 t}; p_{N^2 t})_\infty} \frac{p_t^{-\gamma_j^2/2}}{N} \sum_{\ell=-n}^{n+1} \sin(\gamma_j u_\ell) \theta(-p_t^{1/2} e^{i(x-u_\ell)}; p_t),$$

where $N = N^{C_n}$. The range of summation is now $\{-n, \ldots, n+1\}$ which consists of $2n + 2 = N^{C_n}$ elements. By (1.13),

$$(p_t; p_t)_\infty \theta(-p_t^{1/2} e^{i(x-u_\ell)}; p_t) = \sum_{m \in \mathbb{Z}} p_t^{m^2} e^{i(x-u_\ell)m}.$$

Then we have

$$
L_j^{C_n} = \frac{p_t^{-\gamma_j^2/2}}{(p_{N^2 t}; p_{N^2 t})_\infty} \sum_{m \in \mathbb{Z}} p_t^{m^2/2} e^{ixm}
$$
$$
\times \left\{ \frac{1}{N} \sum_{\ell=-n}^{n+1} e^{i(\gamma-m)u_\ell} - \frac{1}{N} \sum_{\ell=-n}^{n+1} e^{i(-\gamma-m)u_\ell} \right\}.
$$

We can use the identity (4.5) to obtain the equalities

$$
\frac{1}{N} \sum_{\ell=-n}^{n+1} e^{i(\pm\gamma-m)u_\ell} = \sum_{k \in \mathbb{Z}} \mathbf{1}(m = \pm\gamma + Nk) = \sum_{k \in \mathbb{Z}} \mathbf{1}(m = \pm(\gamma + Nk)),
$$

and the above gives

$$
L_j^{C_n} = \frac{p_t^{-\gamma_j^2/2}}{(p_{N^2 t}; p_{N^2 t})_\infty}
$$
$$
\times \left\{ \sum_{k \in \mathbb{Z}} p_t^{(\gamma_j+Nk)^2/2} e^{i(\gamma_j+Nk)x} - \sum_{k \in \mathbb{Z}} p_t^{(\gamma_j+Nk)^2/2} e^{-i(\gamma_j+Nk)x} \right\}.
$$

We see that, for $\gamma_j = \gamma_j^{C_n} = j - n - 1$ and $N = N^{C_n} = 2n + 2$,

$$
\frac{1}{2}(\gamma_j + Nk)^2 = \frac{1}{2}\gamma_j^2 + Njk + N^2 \binom{k}{2}.
$$

Hence we obtain

$$
L_j^{C_n} = \frac{p_t^{-\gamma_j^2/2}}{(p_{N^2 t}; p_{N^2 t})_\infty} \left[p_t^{\gamma_j^2/2} e^{i\gamma x} \sum_{k \in \mathbb{Z}} p_{Nt}^{N\binom{k}{2}} (p_{Nt}^j e^{iNx})^k \right.
$$
$$
\left. - p_t^{\gamma_j^2/2} e^{-i\gamma x} \sum_{k \in \mathbb{Z}} p_{Nt}^{N\binom{k}{2}} (p_{Nt}^j e^{-iNx})^k \right]
$$
$$
= e^{i\gamma x} \theta(-p_{Nt}^j e^{iNx}; p_{Nt}^N) - e^{-i\gamma x} \theta(-p_{Nt}^j e^{-iNx}; p_{Nt}^N).
$$

Since $\gamma_j^{C_n} = j - n - 1 = \alpha_j^{C_n}$ given by (3.15), (4.26) is proved for $R_n = C_n$.
Next consider the case that $R_n = B_n^\vee$. In the second summation in the bracket of (4.27), we change the variable as $-\ell + 1 \to \ell$. Then (4.27) is written as

$$
L_j^{B_n^\vee} = \frac{2i(p_t; p_t)_\infty}{(p_{N^2 t}; p_{N^2 t})_\infty} \frac{p_t^{-\gamma_j^2/2}}{N} \sum_{\ell=-n+1}^{n} \sin(\gamma_j u_\ell) \theta(-p_t^{1/2} e^{i(x-u_\ell)}; p_t),
$$

where the range of summation is $\{-n+1, \ldots, n\}$ consisting of $2n = N^{B_n^\vee}$ elements. The above is written as

$$
L_j^{B_n^\vee} = \frac{p_t^{-\gamma_j^2/2}}{(p_{N^2 t}; p_{N^2 t})_\infty} \sum_{m \in \mathbb{Z}} p_t^{m^2/2} e^{ixm}
$$

$$
\times \left\{ \frac{1}{N} \sum_{\ell=-n+1}^{n} e^{i(\gamma-m)2\pi(\ell-1/2)/N^{B_n^\vee}} - \frac{1}{N} \sum_{\ell=-n+1}^{n} e^{i(-\gamma-m)2\pi(\ell-1/2)/N^{B_n^\vee}} \right\}.
$$

We use the identity (4.5) to obtain the equalities

$$
\frac{1}{N} \sum_{\ell=-n}^{n+1} e^{i(\pm\gamma-m)2\pi(\ell-1/2)/N} = \sum_{k \in \mathbb{Z}} (-1)^k \mathbf{1}(m = \pm(\gamma + Nk))
$$

with $N = N^{B_n^\vee} = 2n$, and then we have

$$
L_j^{B_n^\vee} = \frac{p_t^{-\gamma_j^2/2}}{(p_{N^2 t}; p_{N^2 t})_\infty}
$$

$$
\times \left\{ \sum_{k \in \mathbb{Z}} (-1)^k p_t^{(\gamma_j+Nk)^2/2} e^{i(\gamma_j+Nk)x} - \sum_{k \in \mathbb{Z}} (-1)^k p_t^{(\gamma_j+Nk)^2/2} e^{-i(\gamma_j+Nk)x} \right\}.
$$

Following the calculation shown above for C_n, (4.26) is also proved for $R_n = B_n^\vee$.
(ii) Next we prove (4.26) for $R_n = BC_n$, B_n, and $C_n^\vee$. By (2.26), (4.24), and (4.25), in these cases we have

$$
L_j^{R_n} := \sum_{\ell=1}^{n} \varpi_{j\ell}^{R_n}(p_t) \mathrm{p}_{[0,\pi]}^{\mathrm{ar}}(0, u_\ell; t, x)
$$

$$
= \frac{2i p_t^{1/8}(p_t; p_t)_\infty \, p_t^{-\gamma_j^2/2}}{(p_{N^2 t}; p_{N^2 t})_\infty \, N} \left\{ \sum_{\ell=1}^{n} \widetilde{c}_\ell \sin(\gamma_j u_\ell) e^{-i(x-u_\ell)/2} \theta(-e^{i(x-u_\ell)}; p_t) \right.
$$

$$
\left. - \sum_{\ell=1}^{n} \widetilde{c}_\ell \sin(\gamma_j u_\ell) e^{-i(x+u_\ell)/2} \theta(-e^{i(x+u_\ell)}; p_t) \right\}. \tag{4.28}
$$

Consider the case that $R_n = BC_n$. Since $\sin(\gamma u_\ell^{BC_n}) = \sin(\gamma 2\pi\ell/(2n+1)) = 0$ if $\ell = 0$, we can change the range of ℓ in the first summation in the bracket from $\{1, \ldots, n\}$ to $\{0, 1, \ldots n\}$. In the second summation there, we replace the variable ℓ by $-\ell$ and change the range from $\{1, \ldots, n\}$ to $\{-n, -n+1, \ldots, -1\}$. Since $\widetilde{c}_\ell^{BC_n} = 1 \, \forall \ell$, the above is written as

$$L_j^{BC_n} = \frac{2i p_t^{1/8} (p_t; p_t)_\infty}{(p_{N^2 t}; p_{N^2 t})_\infty} \frac{p_t^{-\gamma_j^2/2}}{N} \sum_{\ell=-n}^{n} \sin(\gamma_j u_\ell) e^{-i(x-u_\ell)/2} \theta(-e^{i(x-u_\ell)}; p_t), \quad (4.29)$$

where the range of summation is $\{-n, \ldots, n\}$ consisting of $2n + 1 = N^{BC_n}$ elements. Next consider the case that $R_n = B_n$. Write the two summations in the bracket of (4.28) as

$$S_+ := \sum_{\ell=1}^{n} \widetilde{c}_\ell \sin(\gamma_j u_\ell) e^{-i(x-u_\ell)/2} \theta(-e^{i(x-u_\ell)}; p_t),$$

$$S_- := -\sum_{\ell=1}^{n} \widetilde{c}_\ell \sin(\gamma_j u_\ell) e^{-i(x+u_\ell)/2} \theta(-e^{i(x+u_\ell)}; p_t). \quad (4.30)$$

If we change the variable of the summations in S_+ and S_- as $-\ell + 1 \to \ell$, we have

$$S_\pm = \mp \sum_{\ell=-n+1}^{0} \widetilde{c}_{-\ell+1} \sin(\gamma_j u_\ell) e^{-i(x\pm u_\ell)/2} \theta(-e^{i(x\pm u_\ell)}; p_t) =: S'_\mp.$$

Therefore,

$$S_+ + S_- = \frac{1}{2}\{(S_+ + S'_+) + (S_- + S'_-)\}$$

$$= \frac{1}{2} \sum_{\ell=-n+1}^{n} \widetilde{c}_\ell \sin(\gamma_j u_\ell) e^{-i(x-u_\ell)/2} \theta(-e^{i(x-u_\ell)}; p_t)$$

$$- \frac{1}{2} \sum_{\ell=-n+1}^{n} \widetilde{c}_\ell \sin(\gamma_j u_\ell) e^{-i(x+u_\ell)/2} \theta(-e^{i(x+u_\ell)}; p_t),$$

where $\widetilde{c}_\ell := \widetilde{c}_{-\ell+1}$ for $\ell = -n + 1, \ldots, 0$. Since $u_\ell^{B_n} = 2\pi(\ell - 1/2)/(2n - 1)$, we see that the term with $\ell = -n + 1$ in the first (resp. second) summation is $-\widetilde{c}_n \sin(\gamma_j \pi) e^{-i(x+\pi)/2} \theta(-e^{i(x+\pi)}; p_t)$ (resp. $\widetilde{c}_n \sin(\gamma_j \pi) e^{-i(x-\pi)/2} \theta(-e^{i(x-\pi)}; p_t)$) and this is equal to the term with $\ell = n$ in the second (resp. first) summation. Since $\widetilde{c}_n^{B_n} = 1/2 = \widetilde{c}_\ell^{B_n}/2, \ell = 1, \ldots, n - 1$, we can conclude that

$$S_+ + S_- = \frac{1}{2} \sum_{\ell=-n+2}^{n} \sin(\gamma_j u_\ell) e^{-i(x-u_\ell)/2} \theta(-e^{i(x-u_\ell)}; p_t)$$

$$- \frac{1}{2} \sum_{\ell=-n+2}^{n} \sin(\gamma_j u_\ell) e^{-i(x+u_\ell)/2} \theta(-e^{i(x+u_\ell)}; p_t),$$

and hence

$$L_j^{B_n} = \frac{ip_t^{1/8}(p_t; p_t)_\infty}{(p_{N^2 t}; p_{N^2 t})_\infty} \frac{p_t^{-\gamma_j^2/2}}{N} \sum_{\ell=-n+2}^{n} \sin(\gamma_j u_\ell)$$

$$\times \left\{ e^{-i(x-u_\ell)/2}\theta(-e^{i(x-u_\ell)}; p_t) - e^{-i(x+u_\ell)/2}\theta(-e^{i(x+u_\ell)}; p_t) \right\}, \qquad (4.31)$$

where the ranges of the two summations are both $\{-n+2, \ldots, n\}$ consisting of $2n - 1 = N^{B_n}$ elements.

Now we consider the case that $R_n = C_n^\vee$. Write the two summations in the bracket of (4.28) as (4.30). Since $\sin(\gamma u_\ell^{C_n^\vee}) = \sin(\gamma_j \pi \ell/n) = 0$ if $\ell = 0$, we can change the ranges of ℓ in S_+ and S_- in the bracket from $\{1, \ldots, n\}$ to $\{0, 1, \ldots, n\}$ with setting $\tilde{c}_0 := 1$. We can also rewrite S_+ and S_- by replacing the variables in the summation as $\ell \to -\ell$. Then we can show that

$$S_+ + S_- = \frac{1}{2} \sum_{\ell=-n}^{n} \tilde{c}_\ell \sin(\gamma_j u_\ell) e^{-i(x-u_\ell)/2}\theta(-e^{i(x-u_\ell)}; p_t)$$

$$- \frac{1}{2} \sum_{\ell=-n}^{n} \tilde{c}_\ell \sin(\gamma_j u_\ell) e^{-i(x+u_\ell)/2}\theta(-e^{i(x+u_\ell)}; p_t),$$

where $\tilde{c}_\ell := \tilde{c}_{-\ell}$ for $\ell = -n, \ldots, -1$. Since $u_\ell^{C_n^\vee} = \pi\ell/n$, we see that the term with $\ell = -n$ in the first (resp. second) summation is $-\tilde{c}_n \sin(\gamma_j \pi) e^{-i(x+\pi)/2}$ $\theta(-e^{i(x+\pi)}; p_t)$ (resp. $\tilde{c}_n \sin(\gamma_j \pi) e^{-i(x-\pi)/2} \theta(-e^{i(x-\pi)}; p_t)$) and this is equal to the term with $\ell = n$ in the second (resp. first) summation. Since $\tilde{c}_n^{C_n^\vee} = 1/2 = \tilde{c}_\ell^{C_n^\vee}/2$, $\ell = 1, \ldots, n - 1$, we can conclude that

$$L_j^{C_n^\vee} = \frac{ip_t^{1/8}(p_t; p_t)_\infty}{(p_{N^2 t}; p_{N^2 t})_\infty} \frac{p_t^{-\gamma_j^2/2}}{N} \sum_{\ell=-n+1}^{n} \sin(\gamma_j u_\ell)$$

$$\times \left\{ e^{-i(x-u_\ell)/2}\theta(-e^{i(x-u_\ell)}; p_t) - e^{-i(x+u_\ell)/2}\theta(-e^{i(x+u_\ell)}; p_t) \right\}, \qquad (4.32)$$

where the ranges of the two summations are both $\{-n+1, \ldots, n\}$ consisting of $2n = N^{C_n^\vee}$ elements. Following the calculations similar to those given in (i), we can verify (4.26) from (4.29), (4.31), and (4.32) for $R_n = BC_n, B_n$, and $C_n^\vee$, respectively. (iii) The proof of (4.26) for $R_n = D_n$ is left for readers as Exercise 4.1. □

Let

$$\det_{1 \le j,k \le n} [\varpi_{jk}^{R_n}(p_t)] =: \begin{cases} i^n \varpi^{R_n}(p_t), & R_n = B_n, B_n^\vee, C_n, C_n^\vee, BC_n, \\ \varpi^{D_n}(p_t), & R_n = D_n. \end{cases} \qquad (4.33)$$

Then $\varpi(p_t) \in \mathbb{R}$ and $\varpi(p_t) \neq 0$ in general by the definition (4.33) with (4.25), and we have the following equalities:

$$\det_{1 \le j,k \le n} [p^{\sharp}_{[0,\pi]}(0, u_j; t, x_k)]$$

$$= \begin{cases} \dfrac{(-i)^n e^{-i \sum_{\ell=1}^n x_\ell/2}}{\varpi(p_t)} \det_{1 \le j,k \le n} [\psi_j^{R_n}(e^{ix_k}; p_{Nt})], & R_n = B_n, C_n^\vee, BC_n, \\[2mm] \dfrac{(-i)^n}{\varpi(p_t)} \det_{1 \le j,k \le n} [\psi_j^{R_n}(e^{ix_k}; p_{Nt})], & R_n = B_n^\vee, C_n, \\[2mm] \dfrac{1}{\varpi(p_t)} \det_{1 \le j,k \le n} [\psi_j^{D_n}(e^{ix_k}; p_{Nt})], & R_n = D_n, \end{cases} \qquad (4.34)$$

$\mathbf{x} \in \mathbb{W}_n([0, \pi])$, where $\mathbf{u} = \mathbf{u}^{R_n}$ given by (4.24), $N = N^{R_n}$ given by (3.9), $\varpi(p_t) = \varpi^{R_n}(p_t)$ given by (4.33), and the boundary conditions are specified as

$$\sharp = \begin{cases} \text{ar,} & \text{for } R_n = B_n, C_n^\vee, BC_n, \\ \text{aa,} & \text{for } R_n = B_n^\vee, C_n, \\ \text{rr,} & \text{for } R_n = D_n. \end{cases} \qquad (4.35)$$

Definition 4.2 Fix $T > 0$. For each $R_n = B_n, B_n^\vee, C_n, C_n^\vee, BC_n, D_n$, the initial and the final configurations are identical and given by $\mathbf{u}^{R_n}$ as (4.24), and the boundary conditions at $x = 0$ and $x = \pi$ are specified by (4.35). Then the system of *noncolliding Brownian bridges* of type R_n is the Markov process $(\mathbf{X}^{R_n}(t))_{t \in [0,T]}$ of n particles on $[0, \pi]$ with time duration T, such that its multi-time joint probability distribution is determined by the following density function: For an arbitrary $m \in \mathbb{N}$ and for an arbitrary series of strictly increasing times $t_0 := 0 < t_1 < \cdots < t_m < T$, the multi-time joint probability density of configurations $\mathbf{x}^{(\ell)} \in \mathbb{W}_n([0, \pi])$ at times $t = t_\ell$, $\ell = 1, \ldots, m$ is given by

$$\mathbf{p}_T^{R_n}(t_1, \mathbf{x}^{(1)}; t_2, \mathbf{x}^{(2)}; \cdots; t_m, \mathbf{x}^{(m)})$$

$$= \prod_{\ell=1}^m \det_{1 \le j,k \le n} [p^{\sharp}_{[0,\pi]}(t_{\ell-1}, x_j^{(\ell-1)}; t_\ell, x_k^{(\ell)})] \frac{p^{\sharp}_{[0,\pi]}(t_m, x_j^{(m)}; T, u_k)}{p^{\sharp}_{[0,\pi]}(0, u_j; T, u_k)}, \qquad (4.36)$$

where $\mathbf{x}^{(0)} := \mathbf{u}^{R_n}$. In particular, the single-time distribution is described by the probability density

$$\mathbf{p}_T^{R_n}(t, \mathbf{x}) = \mathbf{p}_T^{R_n}(0, \mathbf{u}; t, \mathbf{x}; T, \mathbf{u})$$

$$:= \frac{\det_{1 \le j,k \le n} [p^{\sharp}_{[0,\pi]}(0, u_j; t, x_k)] \det_{1 \le j,k \le n} [p^{\sharp}_{[0,\pi]}(t, x_j; T, u_k)]}{\det_{1 \le j,k \le n} [p^{\sharp}_{[0,\pi]}(0, u_j; T, u_k)]}, \qquad (4.37)$$

$t \in [0, T]$, $\mathbf{x} \in \mathbb{W}_n([0, \pi])$.

We can prove the following.

Proposition 4.3 *Let* $R_n = B_n, B_n^\vee, C_n, C_n^\vee, BC_n,$ *and* D_n. *For each* $t \in [0, T]$,

$$
\mathbf{p}_T^{R_n}(t, \mathbf{x}) = \frac{1}{Z^{R_n}} \det_{1 \le j,k \le n} [\psi_j^{R_n}(e^{ix_k}; p_{Nt})] \, \overline{\det_{1 \le j,k \le n} [\psi_j^{R_n}(e^{ix_k}; p_{N(T-t)})]}
$$

$$
= \frac{1}{Z^{R_n}} \det_{1 \le j,k \le n} [\psi_j^{R_n}(e^{ix_k}; p_{Nt})] \, \overline{\det_{1 \le j,k \le n} [\psi_j^{R_n}(e^{ix_k}; p_{N(T-t)})]}, \quad (4.38)
$$

$\mathbf{x} \in \mathbb{W}_n([0, \pi])$, *where* $N = N^{R_n}$ *given by (3.9) and*

$$
Z^{R_n} = Z^{R_n}(p_t, p_T) = \pi^n \prod_{\ell=1}^n h_\ell^{R_n}(p_{Nt}, p_{N(T-t)})
$$

$$
= 2^{n+n_0} \pi^n \sqrt{\theta^{R_n}} \frac{(p_{N^2T}; p_{N^2T})_\infty^n}{(p_{N^2t}; p_{N^2t})_\infty^n (p_{N^2(T-t)}; p_{N^2(T-t)})_\infty^n} \quad (4.39)
$$

with

$$
n_0 = n_0^{R_n} = \begin{cases} 0, & R_n = C_n, C_n^\vee, BC_n, \\ 1, & R_n = B_n, B_n^\vee, \\ 2, & R_n = D_n, \end{cases}
$$

$$
\theta^{R_n} = \theta^{R_n}(p_T) = \begin{cases} \theta(-1; p_{NT})\theta(-1; p_{N^2T}), & R_n = B_n, \\ \theta(-1; p_{NT})\theta(-1; p_{N^2T})/\theta(-p_{N^2T}^{1/2}; p_{N^2T}), & R_n = B_n^\vee, \\ \theta(-1; p_{NT})/\{\theta(-1; p_{N^2T})\theta(-p_{N^2T}^{1/2}; p_{N^2T})\}, & R_n = C_n, \\ \theta(-p_{NT}^{1/2}; p_{NT}), & R_n = C_n^\vee, \\ \theta(-1; p_{NT})/\theta(-1; p_{N^2T}), & R_n = BC_n, \\ \theta(-1; p_{NT})\theta(-1; p_{N^2T})\theta(-p_{N^2T}^{1/2}; p_{N^2T}), & R_n = D_n. \end{cases}
$$

$$(4.40)$$

Proof In (4.34) we replace t by $T - t$ and take the complex conjugate. Notice that by (2.32), $p_{[0,\pi]}^\sharp(0, u_j; T - t, x_k)] = p_{[0,\pi]}^\sharp(t, u_j; T, x_k) = p_{[0,\pi]}^\sharp(t, x_k; T, u_j)$, and that $\varpi(p_t)$ is real-valued by the definition (4.33) with (4.25). We have

$$
\det_{1 \le j,k \le n} [p_{[0,\pi]}^\sharp(t, x_j; T, u_k)]
$$

$$
= \begin{cases} \dfrac{i^n e^{i \sum_{\ell=1}^n x_\ell/2}}{\varpi(p_{T-t})} \det_{1 \le j,k \le n} [\overline{\psi_j^{R_n}(e^{ix_k}; p_{N(T-t)})}], & R_n = B_n, C_n^\vee, BC_n, \\[2.5ex] \dfrac{i^n}{\varpi(p_{T-t})} \det_{1 \le j,k \le n} [\overline{\psi_j^{R_n}(e^{ix_k}; p_{N(T-t)})}], & R_n = B_n^\vee, C_n, \\[2.5ex] \dfrac{1}{\varpi(p_{T-t})} \det_{1 \le j,k \le n} [\overline{\psi_j^{D_n}(e^{ix_k}; p_{N(T-t)})}], & R_n = D_n. \end{cases}
$$

Then the single-time probability density (4.37) is given by

$$\mathbf{p}_T^{R_n}(t, \mathbf{x}) = \frac{1}{Z} \det_{1 \le j,k \le n} [\psi_j^{R_n}(e^{ix_k}; p_{Nt})] \det_{1 \le j,k \le n} \overline{[\psi_j^{R_n}(e^{ix_k}; p_{N(T-t)})]}, \qquad (4.41)$$

where $Z = \varpi(p_t)\varpi(p_{T-t}) \det_{1 \le j,k \le n} [\mathbf{p}_{[0,\pi]}^{\sharp}(0, u_j; T, u_k)] \in \mathbb{R}$. By calculating the integrals of both sides of the equality (4.41) with respect to $\mathbf{x}$ over $\mathbb{W}_n([0, \pi])$, we have the following equalities, where the Andréief identity (4.17) is used:

$$1 = \frac{\pi^n}{Z} \det_{1 \le j,k \le n} \left[\int_{[0,\pi]} \frac{dx}{\pi} \psi^{R_n}(e^{ix}; p_{Nt})\overline{\psi^{R_n}(e^{ix}; p_{N(T-t)})} \right]$$

$$= \frac{\pi^n}{Z} \det_{1 \le j,k \le n} \left[h_j^{R_n}(p_{Nt}, p_{N(T-t)})\delta_{jk} \right]$$

$$= \frac{\pi^n}{Z} \prod_{j=1}^{n} h_j^{R_n}(p_{Nt}, p_{N(T-t)}).$$

Hence $Z = \pi^n \prod_{j=1}^{n} h_j^{R_n}(p_{Nt}, p_{N(T-t)})$. By (3.28) with (3.29) in Proposition 3.4 and the fundamental property of the theta function (1.12), (4.39) with (4.40) is verified. Hence the proof of (4.38) is complete. $\qquad \square$

4.4 Noncolliding Brownian Bridges and Macdonald Denominators

We recall the determinantal identities of Rosengren and Schlosser given by Propositions 3.1 and 3.2 at the beginning of Chap. 3. Then the single-time probability densities of the Brownian bridges given by Propositions 4.2 and 4.3 are written as follows: for $t \in [0, T]$, with $e^{\pm i\mathbf{x}} := (e^{\pm ix_1}, \ldots, e^{\pm ix_n})$,

$$\mathbf{p}_T^{A_{n-1}}(t, \mathbf{x}) = \frac{(p_{nt}; p_{nt})_{\infty}^{n}(p_{n(T-t)}; p_{n(T-t)})_{\infty}^{n}}{(2\pi)^n (p_{n^2 T}; p_{n^2 T})_{\infty}^{n}}$$

$$\times \frac{\theta(r_t e^{i \sum_{\ell=1}^{n} x_\ell}; p_{nt})\theta(r_{T-t}e^{-i \sum_{\ell=1}^{n} x_\ell}; p_{n(T-t)})}{\theta(-|r_T|; p_{nT})}$$

$$\times M^{A_{n-1}}(e^{i\mathbf{x}}; p_{nt})M^{A_{n-1}}(e^{-i\mathbf{x}}; p_{n(T-t)}), \quad \mathbf{x} \in \mathbb{W}_n([0, 2\pi)), \quad (4.42)$$

and

$$
\mathbf{p}_T^{R_n}(t, \mathbf{x}) = \frac{f^{R_n}}{(2\pi)^n \sqrt{\theta^{R_n}}} \frac{(p_{Nt}; p_{Nt})_\infty^n (p_{N(T-t)}; p_{N(T-t)})_\infty^n}{(p_{N^2T}; p_{N^2T})_\infty^n}
$$
$$
\times M^{R_n}(e^{i\mathbf{x}}; p_{Nt}) M^{R_n}(e^{-i\mathbf{x}}; p_{N(T-t)}), \quad \mathbf{x} \in \mathbb{W}_n([0,\pi]), \quad (4.43)
$$

where $N = N^{R_n}$ given by (3.9), $\theta^{R_n} = \theta^{R_n}(p_T)$ given by (4.40), and

$$
f^{R_n} = f^{R_n}(p_t, p_T) = \begin{cases} 2, & R_n = B_n, \\ 2\,(p_{Nt}; p_{Nt}^2)_\infty (p_{N(T-t)}; p_{N(T-t)}^2)_\infty, & R_n = B_n^\vee, \\ 1, & R_n = C_n, BC_n, \\ (p_{Nt}^{1/2}; p_{Nt})_\infty (p_{N(T-t)}^{1/2}; p_{N(T-t)})_\infty, & R_n = C_n^\vee, r \\ 4, & R_n = D_n. \end{cases} \quad (4.44)
$$

Here we take the following limit:

$$
t \to \infty \quad \text{and} \quad T - t \to \infty, \quad (4.45)
$$

Then $p_t \to 0$, $p_{T-t} \to 0$, $r_t \to 0$, $r_{T-t} \to 0$, and we have the following limit distributions, which are independent of time t:

$$
\mathbf{p}^{A_{n-1}}(\mathbf{x}) := \lim_{\substack{t \to \infty, \\ T-t \to \infty}} \mathbf{p}_T^{A_{n-1}}(t, \mathbf{x}) = \frac{1}{(2\pi)^n} |W^{A_{n-1}}(e^{i\mathbf{x}})|^2
$$
$$
= \frac{1}{\widetilde{Z}^{A_{n-1}}} \prod_{1 \le j < k \le n} \sin^2\left(\frac{x_k - x_j}{2}\right), \quad \mathbf{x} \in [0, 2\pi)^n, \quad (4.46)
$$

and for $\mathbf{x} \in [0,\pi]^n$,

$$
\mathbf{p}^{B_n}(\mathbf{x}) := \left. \begin{cases} \lim_{\substack{t\to\infty, \\ T-t\to\infty}} \mathbf{p}_T^{B_n}(t, \mathbf{x}) \\ \lim_{\substack{t\to\infty, \\ T-t\to\infty}} \mathbf{p}_T^{C_n^\vee}(t, \mathbf{x}) \\ \lim_{\substack{t\to\infty, \\ T-t\to\infty}} \mathbf{p}_T^{BC_n}(t, \mathbf{x}) \end{cases} \right\} = \frac{1}{2^n} |W^{B_n}(e^{i\mathbf{x}})|^2
$$
$$
= \frac{1}{\widetilde{Z}^{B_n}} \prod_{1 \le j < k \le n} \left\{ \sin^2\left(\frac{x_k - x_j}{2}\right) \sin^2\left(\frac{x_k + x_j}{2}\right) \right\} \prod_{\ell=1}^n \sin^2\frac{x_\ell}{2}, \quad (4.47)
$$

$$p^{C_n}(\mathbf{x}) := \begin{cases} \lim\limits_{\substack{t \to \infty, \\ T-t \to \infty}} \mathbf{p}_T^{B_n^\vee}(t, \mathbf{x}) \\ \lim\limits_{\substack{t \to \infty, \\ T-t \to \infty}} \mathbf{p}_T^{C_n}(t, \mathbf{x}) \end{cases} = \frac{1}{2^n} |W^{C_n}(e^{i\mathbf{x}})|^2$$

$$= \frac{1}{\widetilde{Z}^{C_n}} \prod_{1 \le j < k \le n} \left\{ \sin^2\left(\frac{x_k - x_j}{2}\right) \sin^2\left(\frac{x_k + x_j}{2}\right) \right\} \prod_{\ell=1}^{n} \sin^2 x_\ell, \tag{4.48}$$

$$p^{D_n}(\mathbf{x}) := \lim_{\substack{t \to \infty, \\ T-t \to \infty}} \mathbf{p}_T^{D_n}(t, \mathbf{x}) = \frac{1}{2^{n+1}} |W^{D_n}(e^{i\mathbf{x}})|^2$$

$$= \frac{1}{\widetilde{Z}^{D_n}} \prod_{1 \le j < k \le n} \left\{ \sin^2\left(\frac{x_k - x_j}{2}\right) \sin^2\left(\frac{x_k + x_j}{2}\right) \right\}. \tag{4.49}$$

We notice that the last factors $\prod_{\ell=1}^{n} \sin^2 \frac{x_\ell}{2}$ in (4.47) and $\prod_{\ell=1}^{n} \sin^2 x_\ell$ in (4.48) are multiplied due to the boundary conditions (4.35) at $x_\ell = 0$ and $x_\ell = \pi$, since $\sin 0 = \sin \pi = 0$, while $\sin(\pi/2) = 1$. Since the boundary condition is rr for the type D_n, such a factor does not exist in (4.49). These temporally homogeneous probability distributions can be regarded as the *trigonometric reductions* of $\mathbf{p}^{R_n}(t, \mathbf{x})$, $R_n = A_{n-1}, B_n, B_n^\vee, C_n, C_n^\vee, BC_n, D_n$, in which the seven types are degenerated to four types. The distribution $P^{A_{n-1}}(d\mathbf{x}) := p^{A_{n-1}}(\mathbf{x})d\mathbf{x}$ given by (4.46) is realized as the eigenvalue distribution of $n \times n$ unitary matrices uniformly distributed in the Haar measure and known as the *circular unitary ensemble* (CUE) in random matrix theory (see, for instance [55, Sect. 11.8]). The distributions $P^{R_n}(d\mathbf{x}) := p^{R_n}(\mathbf{x})d\mathbf{x}$ for $R_n = B_n, C_n$ and D_n are also realized as the eigenvalue distributions of random matrices in the *classical groups*, SO($2n + 1$), Sp(n), and SO($2n$), respectively. See Sect. 2.3c) in [71] and Sect. 5.5 in [16].

Exercise

4.1 Prove (4.26) of Lemma 4.4 for $R_n = D_n$.

Chapter 5
Determinantal Point Processes Associated with Biorthogonal Systems

Abstract The seven types of noncolliding Brownian bridges constructed so far can be regarded as the time-dependent point processes on the interval $[0, 2\pi)$ or $[0, \pi]$ with time duration $[0, T]$. Here we define the correlation functions and their generating function called the characteristic function, which specify the point process. In particular, if all correlation functions are expressed by determinants specified by a two-point continuous function, then the point process is said to be determinantal and the two-point function is called the correlation kernel. We prove that our noncolliding Brownian bridges provide seven types of time-dependent determinantal point processes (DPPs) and their correlation kernels are expressed using the biorthogonal R_n theta functions. We show that if we take the limit $t \to \infty$ and $T - t \to \infty$, the seven time-dependent DPPs are reduced to four types of stationary DPPs. Associated with the diffusive scaling consisting of the proper dilatation and time change, we perform the infinite-particle limit $n \to \infty$. Then we obtain four types of time-dependent DPPs on $\mathbb{R}$ or $\mathbb{R}_+ := [0, \infty)$ with an infinite number of particles with time duration $[0, T]$. Their temporally homogeneous limits are identified with the infinite DPPs well-studied in random matrix theory as the bulk scaling limits of the Gaussian unitary ensemble and its chiral versions. In other words, the present four types of infinite DPPs are elliptic extensions of them.

5.1 Correlation Functions of a Point Process

Let a space S be a subset of $\mathbb{R}^d$ with $d \in \mathbb{N}$ equipped with a *reference measure* λ. A random *point process* with n points, $n \in \mathbb{N}$, on a space S is a statistical ensemble of *nonnegative integer-valued Radon measures*

$$\Xi(\cdot) = \sum_{j=1}^{n} \delta_{X_j}(\cdot),$$

where $\delta_y(\cdot)$, $y \in S$ denotes the *delta measure* (the *Dirac measure*) defined by $\delta_y(\{x\}) = 1$ if $x = y$ and $\delta_y(\{x\}) = 0$ otherwise. In general, the configuration space of a point process is given by

© The Author(s), under exclusive license to Springer Nature Singapore Pte Ltd. 2023   59
M. Katori, *Elliptic Extensions in Statistical and Stochastic Systems*,
SpringerBriefs in Mathematical Physics 47,
https://doi.org/10.1007/978-981-19-9527-9_5

$$\mathrm{Conf}(S) := \left\{ \xi = \sum_i \delta_{x_i} : x_i \in S, \, \xi(\Lambda) < \infty \text{ for all bounded set } \Lambda \subset S \right\}. \quad (5.1)$$

Let $n \in \mathbb{N}$, and consider a point process on S with n particles,[1] $\{X_j\}_{j=1}^n$, which is governed by a probability measure $\mathbf{P}$. We assume that $\mathbf{P}$ has density $\mathbf{p}$ with respect to $\lambda(d\mathbf{x})$. That is, for the set of points $\mathbf{X} := (X_1, \ldots, X_n)$, $\mathbf{P}(\mathbf{X} \in d\mathbf{x}) = \mathbf{p}(\mathbf{x})\lambda(d\mathbf{x})$, $d\mathbf{x} \subset S^n$. Since the labeling order of n points $\{x_j\}_{j=1}^n$ is irrelevant for point configuration $\xi = \sum_{j=1}^n \delta_{x_j}$, the probability density should be normalized as

$$\frac{1}{n!} \int_{S^n} \mathbf{p}(\mathbf{x}) \lambda(d\mathbf{x}) = 1. \quad (5.2)$$

The point process is denoted by a triplet $(\Xi, \mathbf{p}, \lambda(dx))$. For $(\Xi, \mathbf{p}, \lambda(dx))$, the *m-point correlation function*, $1 \le m \le n$, is defined by

$$\rho_m(x_1, \ldots, x_m) = \frac{1}{(n-m)!} \int_{S^{n-m}} \mathbf{p}(x_1, \ldots, x_m, x_{m+1}, \ldots, x_n) \prod_{j=m+1}^n \lambda(dx_j),$$
$$(5.3)$$

$(x_1, \ldots, x_m) \in S^m$. By definition, for any $m \in \{1, \ldots, n\}$, the correlation function ρ_m is a symmetric function on S^m;

$$\rho_m(x_{\sigma(1)}, \ldots, x_{\sigma(m)}) = \rho_m(x_1, \ldots, x_m) \quad \text{for all } \sigma \in \mathfrak{S}_m.$$

Let $\mathcal{B}_c(S)$ be the set of all bounded measurable complex functions with compact support on S. For $\xi \in \mathrm{Conf}(S)$ and $\phi \in \mathcal{B}_c(S)$ we set

$$\langle \xi, \phi \rangle := \int_S \phi(x) \, \xi(dx) = \sum_{j=1}^n \phi(x_j). \quad (5.4)$$

Then the expectation of the random variable $\langle \Xi, \phi \rangle$ with respect to $\mathbf{P}$ is given by

$$\mathbf{E}[\langle \Xi, \phi \rangle] = \int_S \phi(x) \rho_1(x) \lambda(dx).$$

In other words, the *first correlation function* $\rho_1(x)$ gives the *density of points* at $x \in S$ with respect to the reference measure $\lambda(dx)$. With $\xi = \sum_{j=1}^n \delta_{x_j}$, we define $\xi_m := \sum_{i_1, \ldots, i_m : i_j \neq i_k, j \neq k} \delta_{x_{i_1}} \cdots \delta_{x_{i_m}}$, $2 \le m \le n$. Then for all $\phi \in \mathcal{B}_c(S^m)$,

---

[1] In Sects. 5.1–5.3, we only consider point processes with a finite fixed number of points, $n < \infty$. The condition in (5.1) that there is no accumulation of points: $\xi(\Lambda) < \infty$ for all bounded sets $\Lambda \subset S$, should be imposed when we consider infinite particle systems in Sect. 5.4.

$$\mathbf{E}[\langle \Xi_m, \phi \rangle] = \int_{S^m} \phi(x_1, \ldots, x_m) \rho_m(x_1, \ldots, x_m) \prod_{j=1}^{m} \lambda(dx_j).$$

With $\phi \in \mathcal{B}_c(S)$, $\kappa \in \mathbb{R}$, the *characteristic function* of $(\Xi, \mathbf{p}, \lambda(dx))$ is defined by

$$\Psi[\phi; \kappa] := \mathbf{E}\left[e^{\kappa \langle \Xi, \phi \rangle}\right] = \frac{1}{n!} \int_{S^n} e^{\kappa \langle \xi, \phi \rangle} \mathbf{p}(\mathbf{x}) \lambda(d\mathbf{x}), \tag{5.5}$$

which can be regarded as the Laplace transform of the probability density function $\mathbf{p}$. By the normalization (5.2),

$$\Psi[\phi; 0] = 1, \quad \forall \phi \in \mathcal{B}_c(S). \tag{5.6}$$

Put

$$\chi(x) = \chi(x; \kappa) := 1 - e^{\kappa \phi(x)}, \tag{5.7}$$

which we will call a *test function*. By (5.4), we have

$$e^{\kappa \langle \xi, \phi \rangle} = e^{\kappa \sum_{j=1}^{n} \phi(x_j)} = \prod_{j=1}^{n} e^{\kappa \phi(x_j)} = \prod_{j=1}^{n} (1 - \chi(x_j))$$

$$= 1 + \sum_{m=1}^{n} (-1)^m \sum_{1 \le j_1 < j_2 < \cdots < j_m \le n} \prod_{k=1}^{m} \chi(x_{j_k}),$$

where a binomial expansion with respect to χ is performed. Hence

$$\Psi[\phi; \kappa] = 1 + \frac{1}{n!} \sum_{m=1}^{n} (-1)^m \sum_{1 \le j_1 < j_2 < \cdots < j_m \le n} \int_{S^m} \prod_{k=1}^{m} \left\{\chi(x_{j_k}) \lambda(dx_{j_k})\right\}$$

$$\times \int_{S^{n-m}} \prod_{\substack{1 \le \ell \le n, \\ \ell \notin \{j_1, \ldots, j_m\}}} \lambda(dx_\ell) \, \mathbf{p}(\mathbf{x})$$

$$= 1 + \frac{1}{n!} \sum_{m=1}^{n} \frac{n!}{m!(n-m)!} (-1)^m \int_{S^m} \prod_{k=1}^{m} \left\{\chi(x_k) \lambda(dx_k)\right\}$$

$$\times \int_{S^{n-m}} \prod_{\ell=m+1}^{n} \lambda(dx_\ell) \, \mathbf{p}(\mathbf{x}).$$

By the definition (5.3), the above is written as

$$\Psi[\phi; \kappa] = 1 + \sum_{m=1}^{n} (-1)^m \frac{1}{m!} \int_{S^m} \rho_m(x_1, \ldots, x_m) \prod_{k=1}^{m} \left\{\chi(x_k) \lambda(dx_k)\right\}. \tag{5.8}$$

Since $k = 0$ implies $\chi(x) \equiv 0$, the normalization (5.6) is obviously satisfied. This expression means that the characteristic function is regarded as the *generating function of correlation functions*.

5.2 Determinantal Point Processes (DPPs)

If every correlation function is expressed by a determinant in the form

$$\rho_m(x_1, \ldots, x_m) = \det_{1 \le j,k \le m} [K(x_j, x_k)], \quad m = 1, \ldots, n, \tag{5.9}$$

with a two-point continuous function $K(x, y)$, $x, y \in S$, then the point process is said to be a *determinantal point process* (DPP) (or a *fermion point process*) and K is called the *correlation kernel* [68–71] (see also [36, 43] and references therein). Notice that if the total number of points is fixed to be $n \in \mathbb{N}$, then $\operatorname{rank} K = n$. In particular, the density of point with respect to the reference measure λ on S is given by

$$\rho_1(x) = K(x, x), \quad x \in S. \tag{5.10}$$

The characteristic function (5.8) is given by

$$\Psi[\phi; \kappa] = 1 + \sum_{m=1}^{n} (-1)^m \frac{1}{m!} \int_{S^m} \det_{1 \le j,k \le m} [K(x_j, x_k)] \prod_{\ell=1}^{m} \left\{ \chi(x_\ell) \lambda(dx_\ell) \right\}$$

$$= 1 + \sum_{m=1}^{n} (-1)^m \frac{1}{m!} \int_{S^m} \det_{1 \le j,k \le m} [K(x_j, x_k) \chi(x_k)] \prod_{\ell=1}^{m} \lambda(dx_\ell), \tag{5.11}$$

where the multi-linearity of determinant is used. This is regarded as a generating function of determinants and it defines the *Fredholm determinant* written as

$$\Psi[\phi; \kappa] = \operatorname*{Det}_{L^2(S,\lambda)} \left[I - \mathcal{K}\chi \right], \tag{5.12}$$

where I is the identity operator and $\mathcal{K}$ denotes the operator in the square-integrable space $L^2(S, \lambda)$ corresponding to the integral kernel $K(x, y)$, $x, y \in S$ (see, for instance, [43]). We denote the DPP by a triplet $(\Xi, K, \lambda(dx))$.

We will use the following formulas known for determinants [36, 55].

Lemma 5.1 *For fixed $n \in \mathbb{N}$, any permutation $\sigma \in \mathfrak{S}_n$ consists of exclusive cycles. If we write each cyclic permutation as*

$$\mathfrak{c} = \begin{pmatrix} a & b & \cdots & \omega \\ b & c & \cdots & a \end{pmatrix}$$

and the number of cyclic permutations in a given σ as $\ell(\sigma)$, then $\text{sgn}(\sigma) = (-1)^{n-\ell(\sigma)}$. The determinant of $A := (a_{jk})_{1 \le j,k \le n}$ is expressed as

$$\det A = \sum_{\sigma \in \mathfrak{S}_n} (-1)^{n-\ell(\sigma)} \prod_{c_\alpha : 1 \le \alpha \le \ell(\sigma)} \left(a_{ab} a_{bc} \dots a_{wa} \right). \tag{5.13}$$

Lemma 5.2 *For $n \in \mathbb{N}$, let I be an $n \times n$ unit matrix and $A = (a_{jk})_{1 \le j,k \le n}$. The determinant $\det(I - A)$ is expanded as follows:*

$$\det(I - A) = \det_{1 \le j,k \le n} [\delta_{jk} - a_{jk}] = 1 + \sum_{m=1}^{n} \frac{(-1)^m}{m!} D_m[A], \tag{5.14}$$

where

$$D_m[A] := \sum_{j_1=1}^{n} \cdots \sum_{j_m=1}^{n} \det_{1 \le k,\ell \le m} [a_{j_k j_\ell}], \quad m \in \{1, \dots, n\}. \tag{5.15}$$

This is called the Fredholm expansion formula of determinant.

Note that the m-multiple integrals in (5.11) should be regarded as the continuous-variable versions of the m-sums (5.15).

The following useful fact of DPPs is immediately proved using Lemma 5.1.

Lemma 5.3 *Consider a non-vanishing function $f : S \to \mathbb{C}^\times$. Even if the correlation kernel $K(x, y)$ is transformed as*

$$K(x, y) \to K_f(x, y) := f(x) K(x, y) \frac{1}{f(y)}, \quad x, y \in S, \tag{5.16}$$

all correlation functions (5.9) are the same and hence

$$(\Xi, K, \lambda(dx)) \overset{(law)}{=} (\Xi, K_f, \lambda(dx)).$$

The transformation (5.16) is called the *gauge transformation* and the above property of DPP is referred to as *gauge invariance*. See, for instance, Lemma 3.8 in [36].

We can prove the following.

Proposition 5.1 *Fix $n \in \mathbb{N}$. Let $\{f_j\}_{j=1}^n$ and $\{g_j\}_{j=1}^n$ be a pair of functions in $\mathcal{B}_c(S)$ which satisfy the biorthogonality relation,*

$$\int_S f_j(x) g_k(x) \lambda(dx) = h_j \delta_{jk} \quad j, k \in \{1, \dots, n\}, \tag{5.17}$$

where $h_j > 0$, $j = 1, \dots, n$. Assume that the function $\mathbf{p}(\mathbf{x})$ of $\mathbf{x} \in S^n$ given below is nonnegative and provides the probability density function for a point process with respect to $\lambda(dx)$,

$$\mathbf{p}(\mathbf{x}) = \frac{1}{Z} \det_{1 \le j,k \le n} [f_j(x_k)] \det_{1 \le j,k \le n} [g_j(x_k)], \quad \mathbf{x} \in S^n, \qquad (5.18)$$

where $1/Z$ is a normalization factor so that $(1/n!) \int_{S^n} \mathbf{p}(\mathbf{x}) \lambda(d\mathbf{x}) = 1$. Then this point process is a DPP $(\Xi, K, \lambda(dx))$ such that the correlation kernel is given by

$$K(x, y) = \sum_{\ell=1}^{n} \frac{1}{h_\ell} f_\ell(x) g_\ell(y), \quad x, y \in S. \qquad (5.19)$$

Proof For $\phi \in \mathcal{B}_c(S)$, $\kappa \in \mathbb{R}$ with the test function (5.7), the characteristic function of (5.18) is given by

$$
\begin{aligned}
\Psi[\phi; \kappa] &= \frac{1}{n!} \int_{S^n} \prod_{j=1}^{n} (1 - \chi(x_j)) \frac{1}{Z} \det_{1 \le j,k \le n} [f_j(x_k)] \det_{1 \le j,k \le n} [g_j(x_k)] \lambda(d\mathbf{x}) \\
&= \frac{1}{Z} \frac{1}{n!} \int_{S^n} \det_{1 \le j,k \le n} [f_j(x_k)] \det_{1 \le j,k \le n} [(1 - \chi(x_k)) g_j(x_k)] \lambda(d\mathbf{x}) \\
&= \frac{1}{Z} \det_{1 \le j,k \le n} \left[\int_S f_j(x)(1 - \chi(x)) g_k(x) \lambda(dx) \right],
\end{aligned}
$$

where we have used the Andréief identity (4.15). By the normalization condition (5.6), $Z = \det_{1 \le j,k \le n} \left[\int_S f_j(x) g_k(x) \lambda(dx) \right]$, and hence the above is written as

$$
\Psi[\phi; \kappa] = \frac{\det_{1 \le j,k \le n} \left[\int_S f_j(x) g_k(x) \lambda(dx) - \int_S f_j(x) g_k(x) \chi(x) \lambda(dx) \right]}{\det_{1 \le j,k \le n} \left[\int_S f_j(x) g_k(x) \lambda(dx) \right]}.
$$

Define the $n \times n$ matrices $\mathsf{A} := (\mathsf{a}_{jk})_{1 \le j,k \le n}$, $\mathsf{A}^\chi := (\mathsf{a}_{jk}^\chi)_{1 \le j,k \le n}$ as

$$\mathsf{a}_{jk} := \int_S f_j(x) g_k(x) \lambda(dx) = h_j \delta_{jk},$$

$$\mathsf{a}_{jk}^\chi := \int_S f_j(x) g_k(x) \chi(x) \lambda(dx),$$

where the biorthogonality (5.17) was used for a_{jk}. Then we have

$$\Psi[\phi; \kappa] = \frac{\det(\mathsf{A} - \mathsf{A}^\chi)}{\det(\mathsf{A})} = \det(\mathsf{I} - \mathsf{A}^{-1}\mathsf{A}^\chi),$$

where I is the $n \times n$ unit matrix. Here

$$(\mathsf{A}^{-1}\mathsf{A}^{\chi})_{jk} = \sum_{\ell=1}^{n} \frac{1}{h_j}\delta_{j\ell}\int_{S} f_\ell(x)g_k(x)\chi(x)\lambda(dx)$$

$$= \int_{S}\frac{1}{h_j}f_j(x)g_k(x)\chi(x)\lambda(dx).$$

We will apply the Fredholm expansion formula (5.14) in Lemma 5.2. For j_k, $j_\ell \in \{1,\dots,n\}$, $1 \le k,\ell \le m$, $m = 1,\dots,n$, we see that

$$\det_{1\le k,\ell\le m}\left[(\mathsf{A}^{-1}\mathsf{A}^{\chi})_{j_k j_\ell}\right] = \sum_{\sigma\in\mathfrak{S}_m}\mathrm{sgn}(\sigma)\prod_{k=1}^{m}\int_{S}\frac{1}{h_{j_k}}f_{j_k}(x)g_{j_{\sigma(k)}}(x)\chi(x)\lambda(dx)$$

$$= \sum_{\sigma\in\mathfrak{S}_m}\mathrm{sgn}(\sigma)\prod_{k=1}^{m}\int_{S}\frac{1}{h_{j_k}}f_{j_k}(x_{j_k})g_{j_{\sigma(k)}}(x_{j_k})\chi(x_{j_k})\lambda(dx_{j_k})$$

$$= \int_{S^m}\prod_{s=1}^{m}\lambda(dx_{j_s})\sum_{\sigma\in\mathfrak{S}_m}\mathrm{sgn}(\sigma)\prod_{k=1}^{m}\frac{1}{h_{j_k}}f_{j_k}(x_{j_k})g_{j_{\sigma(k)}}(x_{j_k})\chi(x_{j_k})$$

$$= \int_{S^m}\prod_{s=1}^{m}\lambda(dx_{j_s})\det_{1\le k,\ell\le m}\left[\mathsf{b}_{j_k}(x_{j_k})\mathsf{c}_{j_\ell}(x_{j_k})\right],$$

where

$$\mathsf{b}_j(x) := \frac{1}{h_j}f_j(x), \quad \mathsf{c}_j(x) := g_j(x)\chi(x). \tag{5.20}$$

Therefore, we have

$$\Psi[\phi;\kappa] = 1 + \sum_{m=1}^{n}\frac{(-1)^m}{m!}D_m[\mathsf{A}^{-1}\mathsf{A}^{\chi}], \tag{5.21}$$

where

$$D_m[\mathsf{A}^{-1}\mathsf{A}^{\chi}] = \sum_{j_1=1}^{n}\cdots\sum_{j_m=1}^{n}\int_{S^m}\prod_{s=1}^{m}\lambda(dx_{j_s})\det_{1\le k,\ell\le m}\left[\mathsf{b}_{j_k}(x_{j_k})\mathsf{c}_{j_\ell}(x_{j_k})\right]. \tag{5.22}$$

Here we apply the expression (5.13) for determinant given in Lemma 5.1,

$$\det_{1\le k,\ell\le m}\left[\mathsf{b}_{j_k}(x_{j_k})\mathsf{c}_{j_\ell}(x_{j_k})\right] = \sum_{\sigma\in\mathfrak{S}_m}(-1)^{m-\ell(\sigma)}$$

$$\times \prod_{\mathfrak{c}_\alpha:1\le\alpha\le\ell(\sigma)}\left\{(\mathsf{b}_a(x_a)\mathsf{c}_b(x_a))(\mathsf{b}_b(x_b)\mathsf{c}_c(x_b))\cdots(\mathsf{b}_\omega(x_\omega)\mathsf{c}_a(x_\omega))\right\}.$$

Since the product in the bracket is cyclic by definition, we see that

$$
(\mathsf{b}_a(x_a)\mathsf{c}_b(x_a))(\mathsf{b}_b(x_b)\mathsf{c}_c(x_b)) \cdots (\mathsf{b}_\omega(x_\omega)\mathsf{c}_a(x_\omega))
$$
$$
= (\mathsf{c}_b(x_a)\mathsf{b}_b(x_b))(\mathsf{c}_c(x_b)\mathsf{b}_c(x_c)) \cdots (\mathsf{c}_a(x_\omega)\mathsf{b}_a(x_a))
$$

holds. Then (5.22) is written as

$$
D_m[\mathsf{A}^{-1}\mathsf{A}^\chi] = \int_{S^m} \prod_{s=1}^{m} \lambda(dx_s) \sum_{\sigma \in \mathfrak{S}_m} (-1)^{m-\ell(\sigma)}
$$
$$
\times \prod_{c_\alpha : 1 \le \alpha \le \ell(\sigma)} \left[\left(\sum_{b=1}^{n} \mathsf{c}_b(x_a)\mathsf{b}_b(x_b) \right) \left(\sum_{c=1}^{n} \mathsf{c}_c(x_b)\mathsf{b}_c(x_c) \right) \cdots \left(\sum_{a=1}^{n} \mathsf{c}_a(x_\omega)\mathsf{b}_a(x_a) \right) \right]
$$
$$
= \int_{S^m} \det_{1 \le j,k \le m} \left[\sum_{\ell=1}^{n} \mathsf{c}_\ell(x_j)\mathsf{b}_\ell(x_k) \right] \prod_{s=1}^{m} \lambda(dx_s).
$$

Now we can see that (5.21) with (5.22) is written in the form (5.11) with (5.19). Hence the assertion is proved. □

On S, consider a DPP depending on a continuous parameter, or a series of DPPs labeled by a discrete parameter (e.g., the number of points $n \in \mathbb{N}$), and describe the system by $(\Xi_\alpha, K_\alpha, \lambda_\alpha(dx))$ with the continuous or discrete parameter α. We say a point process Ξ_α converges to Ξ as $\alpha \to \infty$ weakly in the *vague topology*, if

$$
\int_S f(x)\Xi_\alpha(dx) \to \int_S f(x)\Xi(dx), \quad \forall f \in C_c(S),
$$

where $C_c(S)$ is the set of all continuous real-valued functions on S with compact support. The *weak convergence* of DPPs is verified by the uniform convergence of the correlation kernel $K_\alpha \to K$ on each compact set $C \subset S \times S$ [69]. We write the weak convergence of DPP as $(\Xi_\alpha, K_\alpha, \lambda_\alpha(dx)) \overset{\alpha \to \infty}{\Longrightarrow} (\Xi, K, \lambda(dx))$.

Applying Proposition 5.1 to the probability measures given by Propositions 4.2 and 4.3, the following are concluded [38].

Theorem 5.1 *(i) Consider the single-time probability measure* $\mathbf{P}_T^{A_{n-1}}(t, d\mathbf{x}) := \mathbf{p}_T^{A_{n-1}}(t, \mathbf{x})d\mathbf{x}$ *for the noncolliding Brownian bridges on* $\mathbb{T}$ *with time duration* T; $t \in [0, T]$. *It provides a time-dependent DPP*

$$
\Xi_T^{A_{n-1}}(t, \cdot) := \sum_{j=1}^{n} \delta_{X_j^{A_{n-1}}(t)}(\cdot), \quad t \in [0, T]
$$

with the uniform measure $\lambda_\mathbb{T}(dx) = dx/2\pi$ *on* $\mathbb{T}$, *such that the correlation kernel is given by*

$$\mathbf{K}_T^{A_{n-1}}(t;x,y) := \sum_{\ell=1}^{n} \frac{\psi_\ell^{A_{n-1}}(e^{ix}; p_{nt}, r_t)\psi_\ell^{A_{n-1}}(e^{-iy}; p_{n(T-t)}, r_{T-t})}{h_\ell^{A_{n-1}}(p_{nt}, p_{n(T-t)}, |r_T|)}, \quad (5.23)$$

$x, y \in [0, 2\pi), t \in [0, T]$.

(ii) For $R_n = B_n$, $B_n^\vee$, C_n, $C_n^\vee$, BC_n, and D_n, consider the single-time probability measures $\mathbf{P}_T^{R_n}(t, d\mathbf{x}) := \mathbf{p}_T^{R_n}(t, \mathbf{x})d\mathbf{x}$ of the noncolliding Brownian bridges on $[0, \pi]$ with time duration T; $t \in [0, T]$. They provide time-dependent DPPs

$$\Xi_T^{R_n}(\cdot) := \sum_{j=1}^{n} \delta_{X_j^{R_n}(t)}(\cdot), \quad t \in [0, T]$$

with the uniform measure $\lambda_{[0,\pi]}(dx) = dx/\pi$ in $[0, \pi]$, such that the correlation kernel is given by

$$\mathbf{K}_T^{R_n}(t;x,y) = \sum_{\ell=1}^{n} \frac{\psi_\ell^{R_n}(e^{ix}; p_{Nt})\psi_\ell^{R_n}(e^{-iy}; p_{N(T-t)})}{h_\ell^{R_n}(p_{Nt}, p_{N(T-t)})}, \quad (5.24)$$

$x, y \in [0, \pi], t \in [0, T]$, where $N = N^{R_n}$ given by (3.9).

Let $\lambda^{A_{n-1}}(dx) = \lambda_{\mathbb{T}}(dx)$ and $\lambda^{R_n}(dx) = \lambda_{[0,\pi]}(dx)$ for $R_n = B_n$, $B_n^\vee$, C_n, $C_n^\vee$, BC_n, D_n. The *time-dependent DPP* of type R_n is the one-parameter $(t \in [0, T])$ family of DPPs, which will be also expressed by a triplet as follows:

$$(\Xi_T^{R_n}, \mathbf{K}_T^{R_n}, \lambda^{R_n}(dx)) := \left((\Xi_T^{R_n}(t, \cdot), \mathbf{K}_T^{R_n}(t;x,y), \lambda^{R_n}(dx))\right)_{t \in [0,T]}. \quad (5.25)$$

5.3 Reductions to Trigonometric DPPs

We take the limit (4.45) considered at the end of Sect. 4.4. By (1.7), we see that

$$\lim_{\substack{t \to \infty, \\ T-t \to \infty}} \psi_\ell^{A_{n-1}}(e^{ix}; p_{nt}, r_t) = e^{i(\ell-1)x},$$

$$\lim_{\substack{t \to \infty, \\ T-t \to \infty}} h_\ell^{A_{n-1}}(p_{nt}, p_{n(T-t)}, |r_T|) = 1, \quad j = 1, \ldots, n.$$

Then it is easy to verify the following:

$$\lim_{\substack{t \to \infty, \\ T-t \to \infty}} \mathbf{K}_T^{A_{n-1}}(t;x,y) = \frac{e^{i(n-1)x/2}}{e^{i(n-1)y/2}} \frac{\sin\{n(x-y)/2\}}{\sin\{(x-y)/2\}}, \quad x, y \in [0, 2\pi).$$

By the gauge invariance of DPP (Proposition 5.3), the factor $e^{i(n-1)x/2}/e^{i(n-1)y/2}$ is irrelevant. We define

$$K^{A_{n-1}}(x, y) := \frac{\sin\{n(x - y)/2\}}{\sin\{(x - y)/2\}}, \quad x, y \in [0, 2\pi), \tag{5.26}$$

then we can say that the point process having the probability density $p^{A_{n-1}}$ given by (4.46) is a DPP and its correlation kernel is given by (5.26). This is realized as the *circular unitary ensemble* (CUE) of eigenvalues of $U(n)$-random matrices [55].

We can also see that

$$\lim_{\substack{t \to \infty, \\ T-t \to \infty}} \psi_\ell^{R_n}(e^{ix}; p_{Nt}) = e^{i\alpha_\ell x} - e^{-i(\alpha_\ell - a)x} = 2i e^{iax/2} \sin\{(\alpha_\ell - a/2)x\},$$

$$\lim_{\substack{t \to \infty, \\ T-t \to \infty}} h_\ell^{R_n}(p_{Nt}, p_{N(T-t)}) = 2, \quad \ell = 1, \dots, n,$$

for $R_n = B_n, B_n^\vee, C_n, C_n^\vee, BC_n$, and

$$\lim_{\substack{t \to \infty, \\ T-t \to \infty}} \psi_\ell^{D_n}(e^{ix}; p_{Nt}) = e^{i\alpha_\ell x} + e^{-i\alpha_\ell x} = 2\cos(\alpha_\ell x),$$

$$\lim_{\substack{t \to \infty, \\ T-t \to \infty}} h_\ell^{D_n}(p_{Nt}, p_{N(T-t)}) = \begin{cases} 2, & \ell = 1, \dots, n - 1, \\ 4, & \ell = n, \end{cases}$$

where $\alpha_\ell = \alpha_\ell^{R_n}$ given by (3.15), $a = a^{R_n}$ given by (3.10), and $\mathcal{N} = \mathcal{N}^{R_n}$ given by (3.9). Hence we have the following kernels expressed by trigonometric functions; for $x, y \in [0, \pi]$:

$$K^{B_n}(x, y) = \begin{cases} \lim_{\substack{t \to \infty, \\ T-t \to \infty}} \mathbf{K}_T^{B_n}(t; x, y) \\ \lim_{\substack{t \to \infty, \\ T-t \to \infty}} \mathbf{K}_T^{C_n^\vee}(t; x, y) \\ \lim_{\substack{t \to \infty, \\ T-t \to \infty}} \mathbf{K}_T^{BC_n}(t; x, y) \end{cases}$$

$$= \frac{1}{2}\left[\frac{\sin\{n(x - y)\}}{\sin\{(x - y)/2\}} - \frac{\sin\{n(x + y)\}}{\sin\{(x + y)/2\}}\right], \tag{5.27}$$

$$K^{C_n}(x, y) = \begin{cases} \lim_{\substack{t \to \infty, \\ T-t \to \infty}} \mathbf{K}_T^{B_n^\vee}(t; x, y) \\ \lim_{\substack{t \to \infty, \\ T-t \to \infty}} \mathbf{K}_T^{C_n}(t; x, y) \end{cases}$$

$$= \frac{1}{2}\left[\frac{\sin\{(2n + 1)(x - y)/2\}}{\sin\{(x - y)/2\}} - \frac{\sin\{(2n + 1)(x + y)/2\}}{\sin\{(x + y)/2\}}\right], \tag{5.28}$$

$$K^{D_n}(x, y) = \lim_{\substack{t \to \infty, \\ T-t \to \infty}} \mathbf{K}_T^{D_n}(t; x, y)$$

$$= \frac{1}{2} \left[\frac{\sin\{(2n-1)(x-y)/2\}}{\sin\{(x-y)/2\}} + \frac{\sin\{(2n-1)(x+y)/2\}}{\sin\{(x+y)/2\}} \right]. \quad (5.29)$$

The DPPs with the above trigonometric correlation kernels K^{B_n}, K^{C_n}, and K^{D_n}, called the *trigonometric DPPs*, are realized by the eigenvalue distributions of random matrices in the *classical groups*, $\mathrm{SO}(2n+1)$, $\mathrm{Sp}(n)$, and $\mathrm{SO}(2n)$, respectively. In other words, the seven types of DPPs $(\Xi_T^{R_n}, \mathbf{K}_T^{R_n}, \lambda^{R_n}(dx))$ given by Theorem 5.1 are the elliptic extensions of these *classical DPPs*.

Remark 5.1 We put emphasis on the fact that the elliptic extensions introduce an additional parameter $t \in [0, T]$ and it is regarded as time, since the t-dependent DPPs (5.25) are nothing but the time sequence of single-time distributions of the noncolliding Brownian bridges. Equilibrium systems can be generalized to *nonequilibrium systems* by elliptic extensions. From such a viewpoint, the above three classifications (5.27)–(5.28) appearing in the temporally homogeneous limit, $t \to \infty, T - t \to \infty$, can be due to the boundary conditions (4.35). That is, the difference of symmetry in classical groups is represented by the difference of boundary conditions imposed on the stochastic processes. In this sense, we can say that the present seven types of non-colliding Brownian bridges $(\mathbf{X}^{R_n}(t))_{t \geq 0}$ and the seven types of time-dependent DPPs $(\Xi_T^{R_n}, \mathbf{K}_T^{R_n}, \lambda^{R_n}(dx))$ are 'stochastic representations' of the seven types of irreducible reduced affine root systems, $R_n = A_{n-1}, B_n, B_n^\vee, C_n, C_n^\vee, BC_n$, and D_n.

5.4 Infinite Particle Systems

5.4.1 Diffusive Scaling Limits

So far we have fixed $n \in \mathbb{N}$ for each system. Here we consider a series of systems with increasing n. For a system of noncolliding Brownian bridges it means increasing the number of paths of Brownian motions, and for a DPP it means increasing the number of points. According to the change of n, we change the scales of spatio-temporal coordinates. Spatial scale change is called *dilatation* and defined as follows for DPPs.

Definition 5.1 For a DPP $(\Xi, K, \lambda(dx))$ with $\Xi = \sum_j \delta_{X_j}$ on a space $S \subset \mathbb{R}^d$, given a factor $c > 0$, we set

$$c \circ \Xi := \sum_j \delta_{cX_j},$$

$$c \circ K(x, y) := K\left(\frac{x}{c}, \frac{y}{c}\right), \quad x, y \in cS := \{cx : x \in S\},$$

$$c \circ \lambda(dx) := c^{-d}\lambda(dx). \quad (5.30)$$

Then the dilatation of the DPP by factor c is defined by $c \circ (\Xi, K, \lambda(dx)) := (c \circ \Xi, c \circ K, c \circ \lambda(dx))$.

Here we notice the following equivalence. For $g \in \mathcal{B}_c(S)$ such that $g : S \to (0, \infty)$,

$$(\Xi, K(x, y), g(x)\lambda(dx)) \overset{\text{(law)}}{=} (\Xi, \sqrt{g(x)}K(x, y)\sqrt{g(y)}, \lambda(dx)). \tag{5.31}$$

In the following we will set

$$\begin{aligned} c &= n \quad \text{for } A_{n-1}, \\ c &= 2n \quad \text{for } B_n, B_n^\vee, C_n, C_n^\vee, BC_n, D_n. \end{aligned} \tag{5.32}$$

For the correlation kernels, associated with the change of spatial variables

$$x \to x/c, \quad y \to y/c, \tag{5.33}$$

we perform the time change as

$$t \to t/c^2, \quad T \to T/c^2. \tag{5.34}$$

The squared factor c^2 in the time change is called the *diffusive scaling*, whose origin is the Brownian motion scaling shown by (2.6) and (2.7) in Sect. 2.1.

We consider the limit $n \to \infty$ in which the spaces S become unbounded as $\mathbb{R}/2\pi c\mathbb{Z} \to \mathbb{R}$ for A_{n-1}, and $[0, \pi c] \to \mathbb{R}_+ := [0, \infty)$ for other R_n, as $c \to \infty$, so that the densities of points remain to be bounded in the limit. Using (5.31), we will set the reference measures on $\mathbb{R}$ and $\mathbb{R}_+$ as the Lebesgue measures on them.

We can prove the following convergence of correlation functions [38].

Proposition 5.2 *For each $t \in [0, T]$, the infinite particle limits of kernels are obtained as follows:*

$$\begin{aligned} \mathcal{K}_T^A(t; x, y) &:= \lim_{n \to \infty} \frac{1}{2\pi n} \mathbf{K}_{T/n^2}^{A_{n-1}}(t/n^2; x/n, y/n) \\ &= \frac{(p_t; p_t)_\infty (p_{T-t}; p_{T-t})_\infty}{2\pi (p_T; p_T)_\infty} \\ &\quad \times \int_0^1 e^{iu(x-y)} \frac{\theta(-p_t^{u+1/2}e^{ix}; p_t)\theta(-p_{T-t}^{u+1/2}e^{-iy}; p_{T-t})}{\theta(-p_T^{u+1/2}; p_T)} du, \\ &\hspace{6cm} x, y \in \mathbb{R}, \end{aligned} \tag{5.35}$$

$$\mathcal{K}_T^B(t; x, y) := \begin{cases} \displaystyle\lim_{n \to \infty} \frac{1}{2\pi n} \mathbf{K}_{T/(2n)^2}^{B_n}(t/(2n)^2; x/2n, y/2n) \\[3mm] \displaystyle\lim_{n \to \infty} \frac{1}{2\pi n} \mathbf{K}_{T/(2n)^2}^{B_n^\vee}(t/(2n)^2; x/2n, y/2n) \end{cases}$$

$$= \frac{(p_t; p_t)_\infty (p_{T-t}; p_{T-t})_\infty}{2\pi (p_T; p_T)_\infty}$$

$$\times \frac{1}{2} \int_{-1}^{1} \left\{ e^{iu(x-y)/2} \frac{\theta(p_t^{(u+1)/2} e^{ix}; p_t) \theta(p_{T-t}^{(u+1)/2} e^{-iy}; p_{T-t})}{\theta(-p_T^{(u+1)/2}; p_T)} \right.$$

$$\left. - e^{iu(x+y)/2} \frac{\theta(p_t^{(u+1)/2} e^{ix}; p_t) \theta(p_{T-t}^{(u+1)/2} e^{iy}; p_{T-t})}{\theta(-p_T^{(u+1)/2}; p_T)} \right\} du,$$

$$x, y \in \mathbb{R}_+, \qquad (5.36)$$

$$\mathcal{K}_T^C(t; x, y) := \begin{cases} \displaystyle\lim_{n\to\infty} \frac{1}{2\pi n} \mathbf{K}_{T/(2n)^2}^{C_n}(t/(2n)^2; x/2n, y/2n) \\[2em] \displaystyle\lim_{n\to\infty} \frac{1}{2\pi n} \mathbf{K}_{T/(2n)^2}^{C_n^\vee}(t/(2n)^2; x/2n, y/2n) \\[2em] \displaystyle\lim_{n\to\infty} \frac{1}{2\pi n} \mathbf{K}_{T/(2n)^2}^{BC_n}(t/(2n)^2; x/2n, y/2n) \end{cases}$$

$$= \frac{(p_t; p_t)_\infty (p_{T-t}; p_{T-t})_\infty}{2\pi (p_T; p_T)_\infty}$$

$$\times \frac{1}{2} \int_{-1}^{1} \left\{ e^{iu(x-y)/2} \frac{\theta(-p_t^{(u+1)/2} e^{ix}; p_t) \theta(-p_{T-t}^{(u+1)/2} e^{-iy}; p_{T-t})}{\theta(-p_T^{(u+1)/2}; p_T)} \right.$$

$$\left. - e^{iu(x+y)/2} \frac{\theta(-p_t^{(u+1)/2} e^{ix}; p_t) \theta(-p_{T-t}^{(u+1)/2} e^{iy}; p_{T-t})}{\theta(-p_T^{(u+1)/2}; p_T)} \right\} du,$$

$$x, y \in \mathbb{R}_+, \qquad (5.37)$$

$$\mathcal{K}_T^D(t; x, y) := \lim_{n\to\infty} \frac{1}{2\pi n} \mathbf{K}_{T/(2n)^2}^{D_n}(t/(2n)^2; x/2n, y/2n)$$

$$= \frac{(p_t; p_t)_\infty (p_{T-t}; p_{T-t})_\infty}{2\pi (p_T; p_T)_\infty}$$

$$\times \frac{1}{2} \int_{-1}^{1} \left\{ e^{iu(x-y)/2} \frac{\theta(-p_t^{(u+1)/2} e^{ix}; p_t) \theta(-p_{T-t}^{(u+1)/2} e^{-iy}; p_{T-t})}{\theta(-p_T^{(u+1)/2}; p_T)} \right.$$

$$\left. + e^{iu(x+y)/2} \frac{\theta(-p_t^{(u+1)/2} e^{ix}; p_t) \theta(-p_{T-t}^{(u+1)/2} e^{iy}; p_{T-t})}{\theta(-p_T^{(u+1)/2}; p_T)} \right\} du,$$

$$x, y \in \mathbb{R}_+. \qquad (5.38)$$

Proof By the diffusive scaling (5.33) and (5.34) with (5.32), the correlation kernel given by (5.23) in Theorem 5.1 is written as

$$\mathbf{K}_{T/n^2}^{A_{n-1}}(t/n^2; x/n, y/n) = \frac{(p_t; p_t)_\infty (p_{T-t}; p_{T-t})_\infty}{(p_T; p_T)_\infty}$$

$$\times \sum_{\ell=1}^{n} e^{i(x-y)(\ell-1)/n} \frac{\theta(-p_t^{(\ell-1)/n}|r_{t/n^2}|e^{ix}; p_t)\theta(-p_{T-t}^{(\ell-1)/n}|r_{(T-t)/n^2}|e^{-iy}; p_{T-t})}{\theta(-p_T^{(\ell-1)/n}|r_{T/n^2}|; p_T)}.$$

By the definition of r_t given by (4.19), we see that $\lim_{n\to\infty} |r_{t/n^2}| = p_t^{1/2}$ and $\lim_{n\to\infty} |r_{(T-t)/n^2}| = p_{T-t}^{1/2}$. Then (5.35) is obtained.

Similarly, if $R_n = B_n, B_n^\vee$, we can see that

$$\lim_{n\to\infty} \frac{1}{2n} \mathbf{K}_{T/(2n)^2}^{R_n}(t/(2n)^2; x/2n, y/2n) = \frac{(p_t; p_t)_\infty (p_{T-t}; p_{T-t})_\infty}{4(p_T; p_T)_\infty}$$

$$\times \int_0^1 \frac{1}{\theta(-p_T^{w/2}; p_T)} \left\{ e^{i(w-1)(x-y)/2} \theta(p_t^{w/2} e^{ix}; p_t) \theta(p_{T-t}^{w/2} e^{-iy}; p_{T-t}) \right.$$

$$+ e^{-i(w-1)(x-y)/2} \theta(p_t^{w/2} e^{-ix}; p_t) \theta(p_{T-t}^{w/2} e^{iy}; p_{T-t})$$

$$- e^{i(w-1)(x+y)/2} \theta(p_t^{w/2} e^{ix}; p_t) \theta(p_{T-t}^{w/2} e^{iy}; p_{T-t})$$

$$\left. - e^{-i(w-1)(x+y)/2} \theta(p_t^{w/2} e^{-ix}; p_t) \theta(p_{T-t}^{w/2} e^{-iy}; p_{T-t}) \right\} dw.$$

We change the integral variable as $w \to u$ by $u = w - 1$. Then $\theta(p_t^{w/2} e^{ix}; p_t) = \theta(p_t^{(u+1)/2} e^{ix}; p_t)$ and $\theta(p_t^{w/2} e^{-ix}; p_t) = \theta(p_t^{(u+1)/2} e^{-ix}; p_t) = \theta(p_t^{(-u+1)/2} e^{ix}; p_t)$, where we have used the symmetry (1.11) of the theta function. Hence the above gives (5.36). We can prove (5.37) and (5.38) in a similar way. □

The above convergences of correlation kernels are uniform in any compact set C in $\mathbb{R} \times \mathbb{R}$ or $\mathbb{R}_+ \times \mathbb{R}_+$ and hence we have the following *weak convergence theorem of DPPs*. Here we observe again the degeneracy [38].

Theorem 5.2 *In the diffusive scaling limit, the seven kinds of time-dependent DPPs with finite numbers of particles converge to the four kinds of time-dependent DPPs with an infinite number of particles; at each time $t \in [0, T]$,*

$$n \circ (\Xi_{T/n^2}^{A_{n-1}}(t/n^2; \cdot), \mathbf{K}_{T/n^2}^{A_{n-1}}(t/n^2; \cdot, \cdot), \lambda_\mathbb{T}(dx)) \overset{n\to\infty}{\Longrightarrow} (\Xi_T^A(t; \cdot), \mathcal{K}_T^A(t; \cdot, \cdot), dx),$$

$$(2n) \circ (\Xi_{T/(2n)^2}^{B_n}(t/(2n)^2; \cdot), \mathbf{K}_{T/(2n)^2}^{B_n}(t/(2n)^2; \cdot, \cdot), \lambda_{[0,\pi]}(dx)) \left.\right\}$$

$$(2n) \circ (\Xi_{T/(2n)^2}^{B_n^\vee}(t/(2n)^2; \cdot), \mathbf{K}_{T/(2n)^2}^{B_n^\vee}(t/(2n)^2; \cdot, \cdot), \lambda_{[0,\pi]}(dx)) \left.\right\}$$

$$\overset{n\to\infty}{\Longrightarrow} (\Xi_T^B(t; \cdot), \mathcal{K}_T^B(t; \cdot, \cdot), \mathbf{1}(x \in \mathbb{R}_+)dx),$$

$$(2n) \circ (\Xi^{C_n}_{T/(2n)^2}(t/(2n)^2; \cdot), \mathbf{K}^{C_n}_{T/(2n)^2}(t/(2n)^2; \cdot, \cdot), \lambda_{[0,\pi]}(dx))$$

$$(2n) \circ (\Xi^{C_n^\vee}_{T/(2n)^2}(t/(2n)^2; \cdot), \mathbf{K}^{C_n^\vee}_{T/(2n)^2}(t/(2n)^2; \cdot, \cdot), \lambda_{[0,\pi]}(dx))$$

$$(2n) \circ (\Xi^{BC_n}_{T/(2n)^2}(t/(2n)^2; \cdot), \mathbf{K}^{BC_n}_{T/(2n)^2}(t/(2n)^2; \cdot, \cdot), \lambda_{[0,\pi]}(dx))$$

$$\overset{n\to\infty}{\Longrightarrow} (\Xi^C_T(t; \cdot), \mathcal{K}^C_T(t; \cdot, \cdot), \mathbf{1}(x \in \mathbb{R}_+)dx),$$

$$(2n) \circ (\Xi^{D_n}_{T/(2n)^2}(t/(2n)^2; \cdot), \mathbf{K}^{D_n}_{T/(2n)^2}(t/(2n)^2; \cdot, \cdot), \lambda_{[0,\pi]}(dx))$$

$$\overset{n\to\infty}{\Longrightarrow} (\Xi^D_T(t; \cdot), \mathcal{K}^D_T(t; \cdot, \cdot), \mathbf{1}(x \in \mathbb{R}_+)dx).$$

In the above, we have observed the degeneracy from the seven series of DPPs with finite numbers of particles to the four types of DPPs with an infinite number of particles.

Remark 5.2 As explained in Sects. 3.1 and 3.2, the R_n theta functions $\{\psi_j^{R_n}\}_{j=1}^n$ of Rosengren and Schlosser are elliptic analogues of 'polynomials of θ with degree n'. The first line of (3.1) in Lemma 3.1 expresses $\psi_j^{A_{n-1}}(e^{ix}; p, r)$ using the theta function θ with the argument proportional to e^{inx} and the nome p^n. For other R_n, (3.17) in Lemma 3.2 gives the expression of $\psi_j^{R_n}(e^{ix}; p)$ using the theta functions with the arguments proportional to $e^{\pm iNx}$ and the nome p^N, where we can see $N \sim 2n$ as $n \to \infty$ from (3.9). Since time t has been introduced by putting $p = p_t := e^{-t}, t \geq 0$ as (4.2) based on the observations in Chap. 2, the diffusive scaling (5.33) and (5.34) with (5.32) does not seem to work in the limit $n \to \infty$, if we only follow the theory of Rosengren and Schlosser [64]. The key ingredients for establishing the diffusive scaling limits are given by (4.7), (4.12) and (4.34), which imply that the KMLGV determinants describing the probability weights of noncolliding paths are proportional to the determinants of $\{\psi_j^{A_{n-1}}(e^{ix_k}; p_{nt}, r_t)\}_{1 \leq j,k \leq n}$ and of $\{\psi_j^{R_n}(e^{ix_k}; p_{Nt})\}_{1 \leq j,k \leq n}$. That is, the time-dependent nome p_t, which we put in defining the KMLGV determinants, should be replaced by p_{nt} or p_{Nt} for the R_n theta functions of Rosengren and Schlosser, when they are used to represent the KMLGV determinants. The formulas for the correlation kernels $\mathbf{K}^{R_n}_T$ given by Theorem 5.1 and their diffusive scaling limits shown by Proposition 5.2 are obtained only if we properly combine the theory of Rosengren and Schlosser [64] with the KMLGV formula as we have shown in Chap. 4. One of the main topics in *nonequilibrium statistical mechanics* is to derive the hydrodynamics from interacting particle systems by taking suitable *scaling limits*. The present results will be regarded as the mathematical realizations of *hydrodynamic limits* for strongly repulsive particle systems.

5.4.2 Temporally Homogeneous Limits

If we take the temporally homogeneous limit (4.45), the four time-dependent DPPs are degenerated to the three stationary DPPs as follows:

$$\lim_{\substack{t \to \infty, \\ T-t \to \infty}} \mathcal{K}_T^A(t; x, y) = \frac{1}{2\pi} \int_0^1 e^{iu(x-y)} du = \frac{\sin\{(x-y)/2\}}{\pi(x-y)}$$

$$=: \mathcal{K}_{\sin}(x, y), \quad x, y \in \mathbb{R}, \tag{5.39}$$

$$\left.\begin{array}{c} \lim_{\substack{t \to \infty, \\ T-t \to \infty}} \mathcal{K}_T^B(t; x, y) \\ \lim_{\substack{t \to \infty, \\ T-t \to \infty}} \mathcal{K}_T^C(t; x, y) \end{array}\right\} = \frac{1}{4\pi} \int_{-1}^1 (e^{iu(x-y)} - e^{iu(x+y)}) du$$

$$= \frac{\sin\{(x-y)/2\}}{\pi(x-y)} - \frac{\sin\{(x+y)/2\}}{\pi(x+y)} =: \mathcal{K}_{\mathrm{chGUE}(1/2)}(x, y), \quad x, y \in \mathbb{R}_+, \tag{5.40}$$

$$\lim_{\substack{t \to \infty, \\ T-t \to \infty}} \mathcal{K}_T^D(t; x, y) = \frac{1}{4\pi} \int_{-1}^1 (e^{iu(x-y)} + e^{iu(x+y)}) du$$

$$= \frac{\sin\{(x-y)/2\}}{\pi(x-y)} + \frac{\sin\{(x+y)/2\}}{\pi(x+y)} =: \mathcal{K}_{\mathrm{chGUE}(-1/2)}(x, y), \quad x, y \in \mathbb{R}_+. \tag{5.41}$$

The limit kernel $\mathcal{K}_{\sin}$ is known as the *sine kernel* or the *sinc kernel*, where the density is given by $\mathcal{K}_{\sin}(x, x) = 1/2\pi$ with respect to the Lebesgue measure on $\mathbb{R}$. This is the correlation kernel of the DPP realized as the bulk scaling limit of the eigenvalue distribution of random matrices in the *Gaussian unitary ensemble* (GUE) [16, 55]. The kernels given by (5.40) and (5.40) are identified with the correlation kernels of the DPPs also studied in random matrix theory. The DPPs with these correlation kernels are realized as the bulk scaling limit of the singular-value distributions of the random matrices in the *chiral Gaussian unitary ensembles* (chGUE) with parameters $v = 1/2$ and $-1/2$, respectively. See [16, 38, 45] and references therein. In other words, we can say that the DPP $(\Xi_T^A(t; \cdot, \cdot), \mathcal{K}_T^A(t; \cdot, \cdot), dx)$ is an elliptic extension of the GUE eigenvalue distribution with the sinc kernel. The other three DPPs, $(\Xi_T^R(t; \cdot, \cdot), \mathcal{K}_T^R(t; \cdot, \cdot), \mathbf{1}(x \in \mathbb{R}_+)dx)$, $R = B, C, D$ given by Theorem 5.2 are elliptic extensions of the chGUE singular-value distributions with $v = \pm 1/2$.

5.4.3 DPPs at Time $t = T/2$

As a special case of the four types of time-dependent DPPs with infinite number of particles, here we study the DPPs at the middle time of duration $t = T/2$. The formulas (5.35)–(5.38) of the correlation kernels are simplified. Moreover, we use *Watson's identity* [57, Sect. 20.7], which is written as

$$\theta(e^{2iz}; p)\theta(e^{2iw}; p) = \frac{(p^2; p^2)_\infty^2}{(p; p)_\infty^2}\Big\{\theta(-e^{2i(z+w)}; p^2)\theta(-pe^{2i(z-w)}; p^2)$$
$$- e^{2iw}\theta(-pe^{2i(z+w)}; p^2)\theta(-e^{2i(z-w)}; p^2)\Big\}, \qquad (5.42)$$

$p \in (0, 1)$, $z, w \in \mathbb{R}$. Then we have the following expressions for the correlation kernels: For $T > 0$,

$$\mathcal{K}_T^A(T/2; x, y) = \frac{(p_T; p_T)_\infty}{2\pi}$$
$$\times \Bigg\{\theta(-p_T^{1/2}e^{i(x+y)}; p_T)\int_0^1 e^{iu(x-y)}\frac{\theta(-p_T^{u+1/2}e^{i(x-y)}; p_T)}{\theta(-p_T^{u+1/2}; p_T)}du$$
$$+ e^{-ix}\theta(-e^{i(x+y)}; p_T)\int_0^1 p_T^{-u/2+1/4}e^{iu(x-y)}\frac{\theta(-p_T^{u}e^{i(x-y)}; p_T)}{\theta(-p_T^{u+1/2}; p_T)}du\Bigg\},$$
$$x, y \in \mathbb{R}, \qquad (5.43)$$

$$\mathcal{K}_T^B(T/2; x, y) = \frac{(p_T; p_T)_\infty}{4\pi}$$
$$\times \Bigg\{\theta(-p_T^{1/2}e^{i(x+y)}; p_T)\int_{-1}^1 e^{iu(x-y)/2}\frac{\theta(-p_T^{(u+1)/2}e^{i(x-y)}; p_T)}{\theta(-p_T^{(u+1)/2}; p_T)}du$$
$$- e^{-ix}\theta(-e^{i(x+y)}; p_T)\int_{-1}^1 p_T^{-u/4+1/4}e^{iu(x-y)/2}\frac{\theta(-p_T^{u/2}e^{i(x-y)}; p_T)}{\theta(-p_T^{(u+1)/2}; p_T)}du$$
$$- \theta(-p_T^{1/2}e^{i(x-y)}; p_T)\int_{-1}^1 e^{iu(x+y)/2}\frac{\theta(-p_T^{(u+1)/2}e^{i(x+y)}; p_T)}{\theta(-p_T^{(u+1)/2}; p_T)}du$$
$$+ e^{-ix}\theta(-e^{i(x-y)}; p_T)\int_{-1}^1 p_T^{-u/4+1/4}e^{iu(x+y)/2}\frac{\theta(-p_T^{u/2}e^{i(x+y)}; p_T)}{\theta(-p_T^{(u+1)/2}; p_T)}du\Bigg\},$$
$$x, y \in \mathbb{R}_+,$$

$$\mathcal{K}_T^C(T/2; x, y) = \frac{(p_T; p_T)_\infty}{4\pi}$$
$$\times \Bigg\{\theta(-p_T^{1/2}e^{i(x+y)}; p_T)\int_{-1}^1 e^{iu(x-y)/2}\frac{\theta(-p_T^{(u+1)/2}e^{i(x-y)}; p_T)}{\theta(-p_T^{(u+1)/2}; p_T)}du$$
$$+ e^{-ix}\theta(-e^{i(x+y)}; p_T)\int_{-1}^1 p_T^{-u/4+1/4}e^{iu(x-y)/2}\frac{\theta(-p_T^{u/2}e^{i(x-y)}; p_T)}{\theta(-p_T^{(u+1)/2}; p_T)}du$$
$$- \theta(-p_T^{1/2}e^{i(x-y)}; p_T)\int_{-1}^1 e^{iu(x+y)/2}\frac{\theta(-p_T^{(u+1)/2}e^{i(x+y)}; p_T)}{\theta(-p_T^{(u+1)/2}; p_T)}du$$
$$- e^{-ix}\theta(-e^{i(x-y)}; p_T)\int_{-1}^1 p_T^{-u/4+1/4}e^{iu(x+y)/2}\frac{\theta(-p_T^{u/2}e^{i(x+y)}; p_T)}{\theta(-p_T^{(u+1)/2}; p_T)}du\Bigg\},$$
$$x, y \in \mathbb{R}_+,$$

$$\mathcal{K}_T^D(T/2; x, y) = \frac{(p_T; p_T)_\infty}{4\pi}$$

$$\times \left\{ \theta(-p_T^{1/2} e^{i(x+y)}; p_T) \int_{-1}^{1} e^{iu(x-y)/2} \frac{\theta(-p_T^{(u+1)/2} e^{i(x-y)}; p_T)}{\theta(-p_T^{(u+1)/2}; p_T)} du \right.$$

$$+ e^{-ix} \theta(-e^{i(x+y)}; p_T) \int_{-1}^{1} p_T^{-u/4+1/4} e^{iu(x-y)/2} \frac{\theta(-p_T^{u/2} e^{i(x-y)}; p_T)}{\theta(-p_T^{(u+1)/2}; p_T)} du$$

$$+ \theta(-p_T^{1/2} e^{i(x-y)}; p_T) \int_{-1}^{1} e^{iu(x+y)/2} \frac{\theta(-p_T^{(u+1)/2} e^{i(x+y)}; p_T)}{\theta(-p_T^{(u+1)/2}; p_T)} du$$

$$\left. + e^{-ix} \theta(-e^{i(x-y)}; p_T) \int_{-1}^{1} p_T^{-u/4+1/4} e^{iu(x+y)/2} \frac{\theta(-p_T^{u/2} e^{i(x+y)}; p_T)}{\theta(-p_T^{(u+1)/2}; p_T)} du \right\},$$

$$x, y \in \mathbb{R}_+.$$

When $y = x$, the first integral in (5.43) becomes 1, and the dependence on coordinates of the second integral disappears. We can use the following evaluation of an integral:

$$\int_0^1 p_T^{-u/2+1/4} \frac{\theta(-p_T^u; p)}{\theta(-p_T^{u+1/2}; p)} du = \frac{1}{T} \mathcal{I}(p_T), \quad T > 0,$$

where

$$\mathcal{I}(p) = \frac{2\pi}{(p; p)_\infty^2} \frac{1}{\theta(-1; p)\theta(-p^{1/2}; p)}.$$

This formula is obtained by first rewriting the ratio of the theta functions in the integrand using the Jacobian elliptic functions and then applying the integral formulas of Jacobian elliptic functions found in, for instance, [57, Sect. 22.14]. Hence we have a much simpler expression for the density function $\rho_T^A(T/2; x) = \mathcal{K}_T^A(T/2; x, x)$, $x \in \mathbb{R}$ as

$$\rho_T^A(T/2, x) = \frac{(p_T; p_T)_\infty}{2\pi} \left\{ \theta(-p_T^{1/2} e^{2ix}; p_T) + \frac{1}{T} \mathcal{I}(p_T) e^{-ix} \theta(-e^{2ix}; p_T) \right\}.$$

$$(5.44)$$

Using a similar formula,

$$\int_{-1}^{1} p_T^{-u/4+1/4} \frac{\theta(-p_T^{u/2}; p_T)}{\theta(-p_T^{(u+1)/2}; p_T)} du = \frac{2}{T} \mathcal{I}(p_T),$$

the density functions $\rho_T^R(T/2; x) = \mathcal{K}_T^R(T/2; x, x), x \in \mathbb{R}_+$ are expressed as follows for $R = B, C, D$:

$$
\rho_T^B(T/2, x) = \frac{(p_T; p_T)_\infty}{2\pi} \left\{ \theta(-p_T^{1/2} e^{2ix}; p_T) - \frac{2}{T} \mathcal{I}(p_T) e^{-ix} \theta(-e^{2ix}; p_T) \right.
$$
$$
- \theta(-p_T^{1/2}; p_T) \int_{-1}^{1} e^{iux} \frac{\theta(-p_T^{(u+1)/2} e^{2ix}; p_T)}{\theta(-p_T^{(u+1)/2}; p_T)} du
$$
$$
\left. + e^{-ix} \theta(-1; p_T) \int_{-1}^{1} p_T^{-u/4+1/4} e^{iux} \frac{\theta(-p_T^{u/2} e^{2ix}; p_T)}{\theta(-p_T^{(u+1)/2}; p_T)} du \right\}, \quad (5.45)
$$

$$
\rho_T^C(T/2, x) = \frac{(p_T; p_T)_\infty}{2\pi} \left\{ \theta(-p_T^{1/2} e^{2ix}; p_T) + \frac{2}{T} \mathcal{I}(p_T) e^{-ix} \theta(-e^{2ix}; p_T) \right.
$$
$$
- \theta(-p_T^{1/2}; p_T) \int_{-1}^{1} e^{iux} \frac{\theta(-p_T^{(u+1)/2} e^{2ix}; p_T)}{\theta(-p_T^{(u+1)/2}; p_T)} du
$$
$$
\left. - e^{-ix} \theta(-1; p_T) \int_{-1}^{1} p_T^{-u/4+1/4} e^{iux} \frac{\theta(-p_T^{u/2} e^{2ix}; p_T)}{\theta(-p_T^{(u+1)/2}; p_T)} du \right\}, \quad (5.46)
$$

$$
\rho_T^D(T/2, x) = \frac{(p_T; p_T)_\infty}{2\pi} \left\{ \theta(-p_T^{1/2} e^{2ix}; p_T) + \frac{2}{T} \mathcal{I}(p_T) e^{-ix} \theta(-e^{2ix}; p_T) \right.
$$
$$
+ \theta(-p_T^{1/2}; p_T) \int_{-1}^{1} e^{iux} \frac{\theta(-p_T^{(u+1)/2} e^{2ix}; p_T)}{\theta(-p_T^{(u+1)/2}; p_T)} du
$$
$$
\left. + e^{-ix} \theta(-1; p_T) \int_{-1}^{1} p_T^{-u/4+1/4} e^{iux} \frac{\theta(-p_T^{u/2} e^{2ix}; p_T)}{\theta(-p_T^{(u+1)/2}; p_T)} du \right\}. \quad (5.47)
$$

It is obvious that

$$
\rho_T^A(T/2; x + 2\pi) = \rho_T^A(T/2; x), \quad \rho_T^A(T/2; -x) = \rho_T^A(T/2; x), \quad x \in \mathbb{R},
$$
$$
\rho_T^C(T/2; 0) = \rho_T^D(T/2; 0) = 0,
$$

and (5.44)–(5.47) give

$$
\lim_{T \to \infty} \rho_T^A(T/2, x) = \frac{1}{2\pi}, \quad x \in \mathbb{R},
$$
$$
\lim_{T \to \infty} \rho_T^B(T/2, x) = \lim_{T \to \infty} \rho_T^C(T/2, x) = \frac{1}{2\pi} \left(1 - \frac{\sin x}{x} \right), \quad x \in \mathbb{R}_+,
$$
$$
\lim_{T \to \infty} \rho_T^D(T/2, x) = \frac{1}{2\pi} \left(1 + \frac{\sin x}{x} \right), \quad x \in \mathbb{R}_+,
$$

which are of course consistent with (5.39)–(5.40).

The numerical plots of the density functions are given in Figs. 5.1–5.4.

Fig. 5.1 The density functions $\rho_T^A(T/2, x)$, $x \in \mathbb{R}$ are plotted for $T = 10$ and $T = 100$ by red and green curves, respectively. The amplitude of oscillation is decreased as T increases and the density function converges to a constant $1/2\pi$ in the limit $T \to \infty$ shown by a blue line

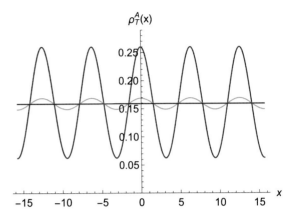

Fig. 5.2 The density functions $\rho_T^B(T/2, x)$, $x \in \mathbb{R}_+$ are plotted for $T = 10$ and $T = 100$ by red and green curves, respectively. In the limit $T \to \infty$, the density function converges to $(1 - \sin x/x)/2\pi$ shown by a blue line

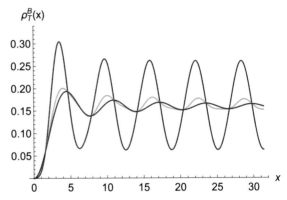

Fig. 5.3 The density functions $\rho_T^C(T/2, x)$, $x \in \mathbb{R}_+$ are plotted for $T = 10$ and $T = 100$ by red and green curves, respectively. In the limit $T \to \infty$, the density function converges to $(1 - \sin x/x)/2\pi$ shown by a blue line

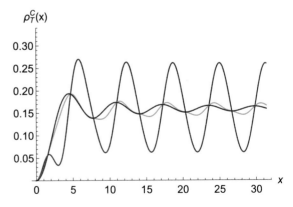

Fig. 5.4 The density functions $\rho_T^D(T/2, x)$, $x \in \mathbb{R}_+$ are plotted for $T = 10$ and $T = 100$ by red and green curves, respectively. In the limit $T \to \infty$, the density function converges to $(1 + \sin x/x)/2\pi$ shown by a blue line

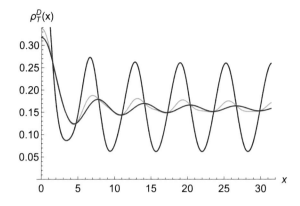

$\rho_T^D(x)$

Exercises

5.1 For $n \in \mathbb{N}$, $T > 0$, we have studied seven types of one-parameter ($t \in [0, T]$) families of n-dimensional spaces of theta functions $\mathcal{E}_{p_{nt}, r_t}^{A_{n-1}}$ and $\mathcal{E}_{p_{Nt}}^{R_n}$, $R_n = B_n, B_n^\vee, C_n$, $C_n^\vee, BC_n, D_n$, where $p_t := e^{-t}$, r_t is given by (4.19), and $\mathcal{N} = \mathcal{N}^{R_n}$ given by (3.9). These function spaces are spanned by the systems of biorthogonal theta functions $\{\psi_j^{A_{n-1}}(e^{ix}; p_{nt}, r_t)\}_{j=1}^n$ defined on $\mathbb{T}$ and $\{\psi_j^{R_n}(e^{ix}; p_{Nt})\}_{j=1}^n$ defined on $[0, \pi]$ for $R_n = B_n, B_n^\vee, C_n, C_n^\vee, BC_n, D_n$. Associated with each system of the biorthogonal theta functions, we have a time-dependent DPP ($\Xi_T^{R_n}, \mathbf{K}_T^{R_n}, \lambda^{R_n}(dx)$) defined by Theorem 5.1. Hence Proposition 5.2 and Theorem 5.2, which give the four types of time-dependent DPPs with an infinite number of particle, suggest four kinds of one-parameter family of infinite-dimensional function spaces $\{\mathcal{E}_{p_t}^R : t \in [0, T]\}$, $R = A, B, C, D$, each of which is spanned by each of the following four systems of theta functions:

$$\psi_u^A(e^{ix}; p_t) := e^{iux}\theta(-p_t^{u+1/2}e^{ix}; p_t), \quad x \in \mathbb{R},$$

$$\psi_u^B(e^{ix}; p_t) := e^{iux/2}\theta(p_t^{(u+1)/2}e^{ix}; p_t) - e^{-iux/2}\theta(p_t^{(u+1)/2}e^{-ix}; p_t) \quad x \in \mathbb{R}_+,$$

$$\psi_u^C(e^{ix}; p_t) := e^{iux/2}\theta(-p_t^{(u+1)/2}e^{ix}; p_t) - e^{-iux/2}\theta(-p_t^{(u+1)/2}e^{-ix}; p_t) \quad x \in \mathbb{R}_+,$$

$$\psi_u^D(e^{ix}; p_t) := e^{iux/2}\theta(-p_t^{(u+1)/2}e^{ix}; p_t) + e^{-iux/2}\theta(-p_t^{(u+1)/2}e^{-ix}; p_t) \quad x \in \mathbb{R}_+,$$

$$\text{with } u \in (0, 1). \tag{5.48}$$

Prove the following.

Proposition 5.3 *For $t \in [0, T]$, the following biorthogonality relations hold:*

$$\int_{\mathbb{R}} \psi_u^A(e^{ix}; p_t)\overline{\psi_w^A(e^{ix}; p_{T-t})}dx = h_u^A(p_t, p_T)\delta(u - w),$$

$$\int_{\mathbb{R}_+} \psi_u^R(e^{ix}; p_t)\overline{\psi_w^R(e^{ix}; p_{T-t})}dx = h_u^R(p_t, p_T)\delta(u - w), \quad R = B, C, D,$$

where

$$h_u^A(p_t, p_T) := \frac{2\pi (p_T; p_T)_\infty}{(p_t; p_t)_\infty (p_{T-t}; p_{T-t})_\infty} \theta(-p_T^{u+1/2}; p_T),$$

$$h_u^R(p_t, p_T) := \frac{2\pi (p_T; p_T)_\infty}{(p_t; p_t)_\infty (p_{T-t}; p_{T-t})_\infty} \theta(-p_T^{(u+1)/2}; p_T), \quad R = B, C, D.$$

Chapter 6
Doubly Periodic Determinantal Point Processes

Abstract We consider the orthogonal R_n theta functions which are the special cases of the biorthogonal functions studied in Chap. 3. Here we extend the real argument x of these functions to complex variable $z = x + iy$ and define seven types of orthonormal R_n theta functions $\{\Psi_j^{R_n}(z)\}_{j=1}^n$ in the fundamental domain $D_{(2\pi, 2\pi|\tau|)}$ in $\mathbb{C}$, which is given by a $2\pi \times 2\pi|\tau|$ rectangular domain. Then seven types of DPPs are introduced so that the correlation functions are expressed by the orthonormal functions. We prove that the probability laws of the obtained DPPs exhibit double periodicity having $D_{(2\pi, 2\pi|\tau|)}$ as the fundamental domain in $\mathbb{C}$. In other words, seven types of DPPs are constructed on the two-dimensional torus $\mathbb{T}^2$. We show that in the infinite-particle limit, the seven types of DPPs on $\mathbb{T}^2$ are reduced to the three types of infinite DPPs on $\mathbb{C}$. One of them is the uniform DPP on $\mathbb{C}$, which is identified with the Ginibre DPP well-studied in random matrix theory. The other two DPPs are new, and are rotationally invariant but inhomogeneous in the radial direction.

6.1 Orthonormal Theta Functions in the Fundamental Domain in $\mathbb{C}$

We consider again the biorthogonal relations shown by Proposition 3.3 in Sect. 3.3 and Proposition 3.4 in Sect. 3.4. The former is for the pairs of A_{n-1} theta functions $\{\psi_j^{A_{n-1}}(e^{ix}; p, r), \psi_j^{A_{n-1}}(e^{ix}; \widehat{p}, \widehat{r})\}_{j=1}^n$ with $p, \widehat{p} \in (0, 1)$, $r, \widehat{r} \in \mathbb{R} \setminus \{0\}$, where the inner product (3.22) is defined by an integral over $x \in [0, 2\pi)$, while the latter is for the pairs of R_n theta functions $\{\psi_j^{R_n}(e^{ix}; p), \psi_j^{R_n}(e^{ix}; \widehat{p})\}_{j=1}^n$ with $p, \widehat{p} \in (0, 1)$, $R_n = B_n, B_n^\vee, C_n, C_n^\vee, BC_n$, and D_n, where the inner product (3.26) is defined by an integral over $x \in [0, \pi]$. Here we simplify the systems by setting $\widehat{p} = p, \widehat{r} = r$, but we replace the real argument x by a complex argument $z = x + iy$, $x, y \in \mathbb{R}$. We can prove the following orthogonality relations with respect to the integral of x over $[0, 2\pi)$.

Lemma 6.1 *Let* $p \in (0, 1)$ *and* $r \in \mathbb{R} \setminus \{0\}$*. For* $j, k = 1, \ldots, n$*, the following orthogonal relations hold:*

© The Author(s), under exclusive license to Springer Nature Singapore Pte Ltd. 2023
M. Katori, *Elliptic Extensions in Statistical and Stochastic Systems*,
SpringerBriefs in Mathematical Physics 47,
https://doi.org/10.1007/978-981-19-9527-9_6

$$\frac{1}{2\pi} \int_0^{2\pi} \psi_j^{A_{n-1}}(e^{i(x+iy)}; p, r)\overline{\psi_k^{A_{n-1}}(e^{i(x+iy)}; p, r)}dx = \widetilde{h}_j^{A_{n-1}}(y; p, r)\delta_{jk}, \quad (6.1)$$

with

$$\widetilde{h}_j^{A_{n-1}}(y; p, r) = e^{-2(j-1)y}\frac{(p^{2n}; p^{2n})_\infty}{(p^n; p^n)_\infty^2}\theta(-r^2 e^{-2ny} p^{2(j-1)}; p^{2n}), \quad (6.2)$$

and for $R_n = B_n$, $B_n^\vee$, C_n, $C_n^\vee$, BC_n, D_n,

$$\frac{1}{2\pi} \int_0^{2\pi} \psi_j^{R_n}(e^{i(x+iy)}; p)\overline{\psi_k^{R_n}(e^{i(x+iy)}; p)}dx = \widetilde{h}_j^{R_n}(y; p)\delta_{jk}, \quad (6.3)$$

with

$$\widetilde{h}_j^{R_n}(y; p) = c_j\frac{(p^{2\mathcal{N}}; p^{2\mathcal{N}})_\infty}{(p^{\mathcal{N}}; p^{\mathcal{N}})_\infty^2}e^{-ay}\left\{e^{-(2j-\delta)y}\theta(-\beta_j(p)^2 e^{-2\mathcal{N}y}; p^{2\mathcal{N}})\right.$$
$$\left. + e^{(2j-\delta)y}\theta(-\beta_j(p)^2 e^{2\mathcal{N}y}; p^{2\mathcal{N}})\right\}, \quad (6.4)$$

where $a = a^{R_n}$ given by (3.10), $\beta_j(p) = \beta_j^{R_n}(p)$ given by (3.16), $\mathcal{N} = \mathcal{N}^{R_n}$ given by (3.9), $c_j = c_j^{R_n}$ given by (3.29), and $\delta = \delta^{R_n}$ given by (3.32), respectively. In particular, if $R_n = B_n$, $B_n^\vee$, D_n,

$$\widetilde{h}_1^{R_n}(y; p) = 4\frac{(p^{2\mathcal{N}}; p^{2\mathcal{N}})_\infty}{(p^{\mathcal{N}}; p^{\mathcal{N}})_\infty^2}e^{(\mathcal{N}-a)y}\theta(-e^{-2\mathcal{N}y}; p^{2\mathcal{N}}), \quad (6.5)$$

and with $\mathcal{N} = \mathcal{N}^{D_n} = 2n - 2$,

$$\widehat{h}_n^{D_n}(y; p) = 4\frac{(p^{2\mathcal{N}}; p^{2\mathcal{N}})_\infty}{(p^{\mathcal{N}}; p^{\mathcal{N}})_\infty^2}\theta(-p^{\mathcal{N}}e^{-2\mathcal{N}y}; p^{2\mathcal{N}}). \quad (6.6)$$

When $y = 0$, the following reduction occurs by definition:

$$\widetilde{h}_j^{A_{n-1}}(0; p, r) = h_j^{A_{n-1}}(p, p, r^2),$$
$$\widehat{h}_j^{R_n}(0; p) = h_j^{R_n}(p, p), \quad j = 1, \ldots, n,$$

for $R_n = B_n$, $B_n^\vee$, C_n, $C_n^\vee$, BC_n, D_n, where $h_j^{A_{n-1}}$ and $h_j^{R_n}$ are given by (3.24) and (3.28).

Proof By the definition of $\psi_j^{A_{n-1}}$ given by (3.1),

$$\psi_j^{A_{n-1}}(e^{i(x+iy)}; p, r) = e^{-(j-1)y}\psi_j^{A_{n-1}}(e^{ix}; p, re^{-ny}),$$

and hence by (3.23) of Proposition 3.3, orthogonality is guaranteed,

$$\frac{1}{2\pi}\int_0^{2\pi}\psi_j^{A_{n-1}}(e^{i(x+iy)};\,p,r)\overline{\psi_k^{A_{n-1}}(e^{i(x+iy)};\,p,r)}dx$$
$$=e^{-2(j-1)y}h_j^{A_{n-1}}(p,\,p,\,r^2e^{-2ny})\delta_{jk}.$$

By (3.24), (6.2) is obtained.

For $R_n=B_n,\,B_n^{\vee},\,C_n,\,C_n^{\vee},\,BC_n,\,D_n$, let $\beta,\,\widehat{\beta}\in\mathbb{R}\setminus\{0\}$, and define the integrals

$$J_{jk,+}^{R_n}(\beta,\widehat{\beta}):=\frac{1}{2\pi}\int_0^{2\pi}e^{i(j-k)x}\theta(\beta e^{i\mathcal{N}x};\,p^{\mathcal{N}})\theta(\widehat{\beta}e^{-i\mathcal{N}x};\,p^{\mathcal{N}})dx,$$

$$J_{jk,-}^{R_n}(\beta,\widehat{\beta}):=\frac{1}{2\pi}\int_0^{2\pi}e^{i(j+k-\delta)x}\theta(\beta e^{i\mathcal{N}x};\,p^{\mathcal{N}})\theta(\widehat{\beta}e^{i\mathcal{N}x};\,p^{\mathcal{N}})dx,$$

$j,k=1,\ldots,n$. The integrals $J_{jk,\pm}^{R_n}$ given by (3.34) in Sect. 3.4 are equal to $J_{jk,\pm}^{R_n}(\beta_j(p),\beta_k(\widehat{p}))$. By the expressions of $\psi_j^{R_n}$ given by (3.17), we can see that

$$\frac{1}{2\pi}\int_0^{2\pi}\psi_j^{R_n}(e^{i(x+iy)};\,p)\overline{\psi_k^{R_n}(e^{i(x+iy)};\,p)}dx$$
$$=\frac{e^{-ay}}{(p^{\mathcal{N}};\,p^{\mathcal{N}})_\infty^2}\Big\{e^{-(j+k-\delta)y}J_{jk,+}^{R_n}(\beta_j(p)e^{-\mathcal{N}y},\,\beta_k(p)e^{-\mathcal{N}y})$$
$$+e^{(j+k-\delta)y}J_{jk,+}^{R_n}(\beta_j(p)e^{\mathcal{N}y},\,\beta_k(p)e^{\mathcal{N}y})$$
$$+\sigma e^{-(j-k)y}J_{jk,-}^{R_n}(\beta_j(p)e^{-\mathcal{N}y},\,\beta_k(p)e^{\mathcal{N}y})$$
$$+\sigma e^{(j-k)y}J_{jk,-}^{R_n}(\beta_j(p)e^{\mathcal{N}y},\,\beta_k(p)e^{-\mathcal{N}y})\Big\}.$$

By a calculation similar to that performed in the proof of Proposition 3.4 with the fact that $\delta^{R_n}-2=\mathcal{N}^{R_n}$ if $R_n=B_n,\,B_n^{\vee},\,D_n$, (6.3) with (6.4) is proved. For (6.5), use the equality $e^{-\mathcal{N}y}\theta(e^{2\mathcal{N}y};\,p^{2\mathcal{N}})=e^{\mathcal{N}y}\theta(e^{-2\mathcal{N}y};\,p^{2\mathcal{N}})$ verified by the inversion formula (1.9) of the theta function. For (6.6), use the fact $\delta^{D_n}=2n$ and the symmetry (1.11) of the theta function. $\qquad\square$

We use the nome modular parameter τ following (2.10). Since we have assumed $p\in(0,1)$, τ is pure imaginary and written as $\tau=i|\tau|$; that is,

$$p=e^{-2\pi|\tau|}.\qquad(6.7)$$

Here we introduce a real parameter $\kappa\in\mathbb{R}$ and represents the norm r of A_{n-1} theta functions as

$$r=(-1)^ne^{-2\pi\kappa|\tau|}=(-1)^np^\kappa.\qquad(6.8)$$

In this setting, the argument of the theta function in (6.2) is written as

$$-r^2 e^{-2ny} p^{2(j-1)} = e^{i\zeta_j},$$

where

$$\zeta_j = 2iny + \eta_j \tag{6.9}$$

with

$$\eta_j = 4\pi i\{(j-1)+\kappa\}|\tau| + \pi. \tag{6.10}$$

Now we apply Jacobi's imaginary transformation (Lemma 2.1) to $\widetilde{h}_j^{A_{n-1}}(y; p, r)$. Here we rewrite the formula (2.12) as follows:

$$\theta(e^{i\zeta}; p) = -ie^{-\zeta^2/4\pi|\tau|+i\zeta/2+\zeta/2|\tau|}$$

$$\times \frac{\widetilde{p}^{1/8}(\widetilde{p}; \widetilde{p})_\infty}{p^{1/8}(p; p)_\infty} \frac{\widetilde{p}}{\sqrt{|\tau|}} \theta(e^{-\zeta/|\tau|}/\widetilde{p}; \widetilde{p}), \tag{6.11}$$

where we have used the fact $\tau = i|\tau|$ and the quasi-periodicity (1.10) of the theta function. Notice that the nome of the theta function in (6.2) is p^{2n}. Hence we have to use the formula (6.11), replacing $p \to p^{2n} = e^{-4\pi n|\tau|}$, $\widetilde{p} \to \widetilde{p^{2n}} = e^{-\pi/n|\tau|}$, and $|\tau| \to 2n|\tau|$. By (6.9) with (6.10), we can see that

$$-\frac{\zeta_j^2}{8n\pi|\tau|} + \frac{i\zeta_j}{2} + \frac{\zeta_j}{4n|\tau|} = \frac{ny^2}{2\pi|\tau|} + \{2(j-1)+(2\kappa-n)\}y - \frac{\eta_j^2}{8n\pi|\tau|} + \frac{i\eta_j}{2} + \frac{\eta_j}{4n|\tau|}.$$

Moreover, by (6.10), we can show that

$$-\frac{\eta_j^2}{8n\pi|\tau|} + \frac{i\eta_j}{2} + \frac{\eta_j}{4n|\tau|} = \frac{2\pi|\tau|}{n}\left(j-1+\kappa-\frac{n}{2}\right)^2 - \frac{\pi|\tau|n}{2} + \frac{i\pi}{2} + \frac{\pi}{8n|\tau|}.$$

Since $e^{i\pi/2} = i$, $e^{-\pi|\tau|n/2} = (p^{2n})^{1/8}$, and $e^{\pi/8n|\tau|} = (\widetilde{p^{2n}})^{-1/8}$, we see that

$$-ie^{-\zeta_j^2/8n\pi|\tau|+i\zeta_j/2+\zeta_j/4n|\tau|}$$

$$= e^{ny^2/2\pi|\tau|+\{2(j-1)+(2\kappa-n)\}y} e^{2\pi|\tau|(j-1+\kappa-n/2)^2/n} \frac{(p^{2n})^{1/8}}{(\widetilde{p^{2n}})^{1/8}}.$$

Hence we have

$$\widetilde{h}_j^{A_{n-1}}(y; e^{-2\pi|\tau|}, (-1)^n e^{-2\pi\kappa})$$

$$= e^{-2(j-1)y} \frac{(p^{2n}; p^{2n})_\infty}{(p^n; p^n)_\infty^2}$$

$$\times\, e^{ny^2/2\pi|\tau|+\{2(j-1)+(2\kappa-n)\}y}\, e^{2\pi|\tau|(j-1+\kappa-n/2)^2/n}\, \frac{(\widetilde{p^{2n}})^{1/8}}{(\widetilde{p^{2n}})^{1/8}}$$

$$\times\, \frac{(\widetilde{p^{2n}})^{1/8}}{(\widetilde{p^{2n}})^{1/8}}\, \frac{(\widetilde{p^{2n}};\,\widetilde{p^{2n}})_\infty}{(\widetilde{p^{2n}};\,\widetilde{p^{2n}})_\infty}\, \frac{\widetilde{p^{2n}}}{\sqrt{2n|\tau|}}\, \theta(e^{-\zeta_j/2n|\tau|}/\widetilde{p^{2n}};\,\widetilde{p^{2n}})$$

$$=\, \frac{1}{\sqrt{2n|\tau|}}\, \frac{\widetilde{p^{2n}}(\widetilde{p^{2n}};\,\widetilde{p^{2n}})_\infty}{(p^n;\,p^n)_\infty^2}\, e^{2\pi|\tau|(j-1+\kappa-n/2)^2/n+ny^2/2\pi|\tau|+(2\kappa-n)y}$$

$$\times\, \theta(e^{-iy/|\tau|-2\pi i(j-1+\kappa)/n+\pi/2n|\tau|};\,\widetilde{p^{2n}}).$$

The above result is summarized as follows:

Lemma 6.2 *Assume (6.7) and (6.8). Then (6.2) is written as follows:*

$$\widetilde{h}_j^{A_{n-1}}(y;\,p,r) = C_1(n|\tau|)(e^{-\pi/n|\tau|};\,e^{-\pi/n|\tau|})_\infty$$

$$\times\, e^{2\pi|\tau|(j-1+\kappa-n/2)^2/n+ny^2/2\pi|\tau|+(2\kappa-n)y}$$

$$\times\, \theta(C_2(n|\tau|,(j-1+\kappa)|\tau|)e^{-iy/|\tau|};\,e^{-\pi/n|\tau|}), \tag{6.12}$$

$j = 1,\ldots,n$, *where*

$$C_1(t) := \frac{1}{\sqrt{2t}}\, \frac{e^{-\pi/t}}{(e^{-2\pi t};\,e^{-2\pi t})_\infty^2}, \qquad C_2(t,s) := e^{-(4is-1)\pi/2t}. \tag{6.13}$$

In particular, if $\kappa = n/2$, then

$$r = (-1)^n e^{-n\pi|\tau|} = (-1)^n p^{n/2}. \tag{6.14}$$

For $j = 1,\ldots,n$, we have

$$\widetilde{h}_j^{A_{n-1}}(y;\,p,r) = C_1(n|\tau|)(e^{-\pi/n|\tau|};\,e^{-\pi/n|\tau|})_\infty e^{2\pi|\tau|(j-1)^2/n+ny^2/2\pi|\tau|}$$

$$\times\, \theta(C_2(n|\tau|,(j-1+n/2)|\tau|)e^{-iy/|\tau|};\,e^{-\pi/n|\tau|}). \tag{6.15}$$

In order to present similar expressions for $\widetilde{h}^{R_n}(y;\,p)$ given by (6.4) for $R_n = B_n$, $B_n^\vee$, C_n, $C_n^\vee$, BC_n, D_n, we introduce the notation

$$J(j) = J^{R_n}(j) := \alpha_j - a/2 = \begin{cases} j - n - 1/2, & R_n = B_n, C_n^\vee, BC_n, \\ j - n - 1, & R_n = B_n^\vee, C_n, \\ j - n, & R_n = D_n, \end{cases} \tag{6.16}$$

where $a = a^{R_n}$ given by (3.10) and $\alpha_j = \alpha_j^{R_n}$ given by (3.15).

Lemma 6.3 *Assume (6.7). Then for $R_n = B_n$, $B_n^\vee$, C_n, $C_n^\vee$, BC_n, D_n, (6.4) is written as follows:*

$$\widetilde{h}_j^{R_n}(y; p) = c_j C_1(\mathcal{N}|\tau|)(e^{-\pi/\mathcal{N}|\tau|}; e^{-\pi/\mathcal{N}|\tau|})_\infty e^{2\pi|\tau|J(j)^2/\mathcal{N} + \mathcal{N}y^2/2\pi|\tau| - ay}$$

$$\times \left\{ \theta(C_2(\mathcal{N}|\tau|, J(j)|\tau|)e^{-iy/|\tau|}; e^{-\pi/\mathcal{N}|\tau|}) \right.$$

$$\left. + \theta(C_2(\mathcal{N}|\tau|, J(j)|\tau|)e^{iy/|\tau|}; e^{-\pi/\mathcal{N}|\tau|}) \right\}, \quad j = 1, \dots, n, \quad (6.17)$$

where $a = a^{R_n}$ given by (3.10), $C_1(t)$ and $C_2(t, s)$ are defined by (6.13), $c_j = c_j^{R_n}$ given by (3.29), $\mathcal{N} = \mathcal{N}^{R_n}$ given by (3.9), and $J(j) = J^{R_n}(j)$ given by (6.16).

Proof Notice that the following hold for any $R_n = B_n, B_n^\vee, C_n, C_n^\vee, BC_n, D_n$:

$$2j - \delta^{R_n} = 2J^{R_n}(j), \quad j = 1, \dots, n.$$

Then a calculation similar to that given for Lemma 6.2 proves the present lemma. □

For (6.15) in Lemma 6.2 and (6.17) in Lemma 6.3, it is easy to verify the following.

Lemma 6.4 *With (6.7), we have*

$$\frac{1}{2\pi|\tau|} \int_0^{2\pi|\tau|} e^{-ny^2/2\pi|\tau|} \widetilde{h}_j^{A_{n-1}}(y; p, (-1)^n p^{n/2}) dy = C_1(n|\tau|)e^{2\pi|\tau|(j-1)^2/n},$$

$$\frac{1}{2\pi|\tau|} \int_0^{2\pi|\tau|} e^{-\mathcal{N}y^2/2\pi|\tau| + ay} \widetilde{h}_j^{R_n}(y; p) dy = 2c_j C_1(\mathcal{N}|\tau|)e^{2\pi|\tau|J(j)^2/\mathcal{N}},$$

$j = 1, \dots, n$, for $R_n = B_n, B_n^\vee, C_n, C_n^\vee, BC_n, D_n$, where $\mathcal{N} = \mathcal{N}^{R_n}$ given by (3.9), $a = a^{R_n}$ given by (3.10), $c_j = c_j^{R_n}$ given by (3.29), and $J(j) = J^{R_n}(j)$ given by (6.16).

Proof The assertions are immediately obtained, if we notice the following. Assume that C does not depend on y. Then

$$\frac{1}{2\pi|\tau|} \int_0^{2\pi|\tau|} \theta(Ce^{\pm iy/|\tau|}; p) dy$$

$$= \frac{1}{(p; p)_\infty} \sum_{k \in \mathbb{Z}} (-1)^k p^{\binom{k}{2}} C^k \times \frac{1}{2\pi|\tau|} \int_0^{2\pi|\tau|} e^{\pm iyk/|\tau|} dy$$

$$= \frac{1}{(p; p)_\infty} \sum_{k \in \mathbb{Z}} (-1)^k p^{\binom{k}{2}} C^k \delta_{k0} = \frac{1}{(p; p)_\infty}.$$

Hence the assertions are proved.   □

With (6.7), define

$$\Psi_j^{A_{n-1}}(z) = \Psi_j^{A_{n-1}}(z; |\tau|)$$

$$:= \frac{e^{-\pi|\tau|(j-1)^2/n}}{\sqrt{C_1(n|\tau|)}} e^{-ny^2/4\pi|\tau|} \psi_j^{A_{n-1}}(e^{iz}; p, (-1)^n p^{n/2}), \quad (6.18)$$

and for $R_n = B_n, B_n^\vee, C_n, C_n^\vee, BC_n, D_n,$

$$\Psi_j^{R_n}(z) = \Psi_j^{R_n}(z; |\tau|)$$

$$:= \frac{e^{-\pi|\tau|J(j)^2/\mathcal{N}}}{\sqrt{2c_j C_1(\mathcal{N}|\tau|)}} e^{-\mathcal{N}y^2/4\pi|\tau|+ay/2} \psi_j^{R_n}(e^{iz}; p), \qquad (6.19)$$

$j = 1, \ldots, n$, $z = x + iy \in \mathbb{C}$, where $\mathcal{N} = \mathcal{N}^{R_n}$ given by (3.9), $a = a^{R_n}$ given by (3.10), $c_j = c_j^{R_n}$ given by (3.29), C_1 is given by (6.13), and $J(j) = J^{R_n}(j)$ given by (6.16). We consider a rectangular domain in $\mathbb{C}$,

$$D_{(2\pi, 2\pi|\tau|)} := \{z = x + iy \in \mathbb{C} : 0 \le x < 2\pi, 0 \le y < 2\pi|\tau|\}. \qquad (6.20)$$

The area of this domain is given by $|D_{(2\pi, 2\pi|\tau|)}| = 4\pi^2|\tau|$. We introduce the inner product for holomorphic functions f, g defined on $D_{(2\pi, 2\pi|\tau|)}$ by

$$\langle f, g \rangle_{D_{(2\pi, 2\pi|\tau|)}} := \frac{1}{|D_{(2\pi, 2\pi|\tau|)}|} \int_{D_{(2\pi, 2\pi|\tau|)}} f(z)\overline{g(z)} dz$$

$$= \frac{1}{(2\pi)^2|\tau|} \int_0^{2\pi} dx \int_0^{2\pi|\tau|} dy\, f(x + iy)\overline{g(x + iy)}. \qquad (6.21)$$

The above results are summarized as follows [39].

Proposition 6.1 *For* $R_n = A_{n-1}, B_n, B_n^\vee, C_n, C_n^\vee, BC_n, D_n,$

$$\langle \Psi_j^{R_n}, \Psi_k^{R_n} \rangle_{D_{(2\pi, 2\pi|\tau|)}} = \delta_{jk}, \quad j, k = 1, \ldots, n. \qquad (6.22)$$

The orthonormal functions $\{\Psi_j^{R_n}\}_{j=1}^n$ have the following *doubly-quasi-periodicity*. Hence the rectangular domain (6.20) can be regarded as the *fundamental domain* in $\mathbb{C}$ and $\{\Psi_j^{R_n}\}_{j=1}^n$ will be extended as the holomorphic functions on $\mathbb{C}$.

Lemma 6.5 *The following equalities hold, where* $R_n = B_n, B_n^\vee, C_n, C_n^\vee, BC_n, D_n$:

$$\Psi_j^{A_{n-1}}(z + 2\pi) = \Psi_j^{A_{n-1}}(z), \ \Psi_j^{R_n}(z + 2\pi) = \Psi_j^{R_n}(z), \qquad (6.23)$$

$$\Psi_j^{A_{n-1}}(z + 2\pi i|\tau|) = e^{-inx} \Psi_j^{A_{n-1}}(z), \qquad (6.24)$$

$$\Psi_j^{R_n}(z + 2\pi i|\tau|) = \sigma_1 e^{-i\mathcal{N}x} \Psi_j^{R_n}(z), \qquad (6.25)$$

$$\Psi_j^{R_n}(-z) = \sigma_2 e^{-iax} \Psi_j^{R_n}(z), \qquad (6.26)$$

$j = 1, \ldots, n$, *where* $\mathcal{N} = \mathcal{N}^{R_n}$ *given by (3.9),* $a = a^{R_n}$ *given by (3.10),* $\sigma_1 = \sigma_1^{R_n}$ *given by (3.11), and* $\sigma_2 = \sigma_2^{R_n}$ *given by (3.12) for* $R_n = B_n, B_n^\vee, C_n, C_n^\vee, BC_n, D_n$.

Proof The periodicity (6.23) is obvious, since $\{\Psi_j^{R_n}(z)\}_{j=1}^n$ are functions of e^{iz} as shown by (6.18) and (6.19). Since $\psi_j^{A_{n-1}}(e^{iz}; p, (-1)^n p^{n/2}) \in \mathcal{E}_{p,(-1)^n p^{n/2}}^{A_{n-1}}$ with (6.7),

Definition 3.1 gives

$$\psi_j^{A_{n-1}}(e^{i(z+2\pi|\tau|i)};\, p,\, (-1)^n p^{n/2}) = \psi_j^{A_{n-1}}(pe^{iz};\, p,\, (-1)^n p^{n/2})$$

$$= \frac{(-1)^n \psi_j^{A_{n-1}}(e^{iz};\, p,\, (-1)^n p^{n/2})}{(-1)^n p^{n/2}(e^{iz})^n} = e^{n\pi|\tau|}e^{ny}e^{-inx}\psi_j^{A_{n-1}}(e^{iz};\, p,\, (-1)^n p^{n/2}).$$

Similarly, since $\psi_j^{R_n}(e^{iz};\, p) \in \mathcal{E}_p^{R_n}$ with (6.7), Definition 3.3 with (3.13) gives

$$\psi_j^{R_n}(e^{i(z+2\pi|\tau|i)};\, p) = \psi_j^{R_n}(pe^{iz};\, p)$$

$$= \frac{\sigma_1 \psi_j^{R_n}(e^{iz};\, p)}{p^{(\mathcal{N}-a)/2}(e^{iz})^{\mathcal{N}}} = \sigma_1 e^{(\mathcal{N}-a)\pi|\tau|}e^{\mathcal{N}y}e^{-i\mathcal{N}x}\psi_j^{R_n}(e^{iz};\, p),$$

for $R_n = B_n,\, B_n^{\vee},\, C_n,\, C_n^{\vee},\, BC_n,\, D_n$. We can see that

$$e^{-n(y+2\pi|\tau|)^2/4\pi|\tau|} = e^{-n\pi|\tau|}e^{-ny}e^{-ny^2/4\pi|\tau|},$$

$$e^{-\mathcal{N}(y+2\pi|\tau|)^2/4\pi|\tau|+a(y+2\pi|\tau|)/2} = e^{-(\mathcal{N}-a)\pi|\tau|}e^{-\mathcal{N}y}e^{-\mathcal{N}y^2/4\pi|\tau|+ay/2}.$$

Hence the quasi-periodicities (6.24) and (6.25) are proved. By (3.14),

$$\psi^{R_n}(e^{-iz};\, p) = \psi^{R_n}(1/e^{iz};\, p) = \sigma_2 e^{-iaz}\psi^{R_n}(e^{iz};\, p) = e^{ay}\sigma_2 a e^{-iax}\psi^{R_n}(e^{iz};\, p).$$

Since $e^{-\mathcal{N}(-y)^2/4\pi|\tau|+a(-y)/2} = e^{-ay}e^{-\mathcal{N}y^2/4\pi|\tau|+ay/2}$, the inversion formula (6.26) is proved.   □

6.2 DPPs on the Two-Dimensional Torus $\mathbb{T}^2$

For $R_n = A_{n-1},\, B_n,\, B_n^{\vee},\, C_n,\, C_n^{\vee},\, BC_n,\, D_n$, we put

$$k^{R_n}(z, z') := \sum_{j=1}^{n} \Psi_j^{R_n}(z)\overline{\Psi_j^{R_n}(z')}, \quad z, z' \in D_{(2\pi, 2\pi|\tau|)}.$$

By Lemma 6.5, the following *double periodicity* (modulo gauge transformation) is obvious:

$$k^{R_n}(z + 2\pi, z' + 2\pi) = k^{R_n}(z, z'),$$

$$k^{R_n}(z + 2\pi|\tau|i, z' + 2\pi|\tau|i) = \frac{e^{-i\mathcal{N}x}}{e^{-i\mathcal{N}x'}}k^{R_n}(z, z'), \quad z, z' \in \mathbb{C}, \tag{6.27}$$

Fig. 6.1 By imposing double periodicity, the rectangular domain $D_{(2\pi, 2\pi|\tau|)}$ in $\mathbb{C}$ is identified with the two-dimensional torus $\mathbb{T}^2$

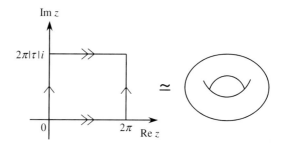

where for $R_n = A_{n-1}$, we define $\mathcal{N} = \mathcal{N}^{R_n} = n$ and for other R_n, $\mathcal{N} = \mathcal{N}^{R_n}$ given by (3.9). As a matter of course, orthogonality is a special case of biorthogonality; the argument given in Sect. 5.2 implies that we have obtained seven types of DPPs on $\mathbb{C}$ whose probability laws have double periodicity such that the *fundamental domain* is given by $D_{(2\pi, 2\pi|\tau|)}$ defined by (6.20). If we define the two-dimensional torus by

$$\mathbb{T}^2 = \mathbb{T}^2_{|\tau|} := \{z \in \mathbb{C} : z + 2\pi = z, z + 2\pi|\tau|i = z\}$$
$$\simeq (\mathbb{R}/2\pi\mathbb{Z}) \times (\mathbb{R}/2\pi|\tau|\mathbb{Z}),$$

then we can also say that we have seven types of DPPs on $\mathbb{T}^2$ [39]. See Fig. 6.1.

Theorem 6.1 *The seven types of point processes*

$$\Xi^{R_n}_{\mathbb{T}^2}(\cdot) = \sum_{j=1}^{n} \delta_{X^{R_n}_j}(\cdot),$$

for $R_n = A_{n-1}, B_n, B_n^{\vee}, C_n, C_n^{\vee}, BC_n, D_n$ are well defined on $\mathbb{T}^2$ so that they have the probability densities $\mathbf{p}^{R_n}_{\mathbb{T}^2}(\mathbf{z})$,

$$\mathbf{p}^{R_n}_{\mathbb{T}^2}(\mathbf{z}) = \frac{1}{Z^{R_n}} \left| \det_{1 \leq j,k \leq n} [\Psi^{R_n}_j(z_k)] \right|^2, \quad \mathbf{z} = (z_1, \ldots, z_n) \in (\mathbb{T}^2)^n,$$

with respect to the Lebesgue measure $d\mathbf{z} := \prod_{j=1}^{n} d\mathrm{Re}\, z_j d\mathrm{Im}\, z_j$ on $\mathbb{C}^n$. Here $1/Z^{R_n}$ are the normalization factors such that $(1/n!) \int_{(\mathbb{T}^2)^n} \mathbf{p}^{R_n}_{\mathbb{T}^2}(\mathbf{z})d\mathbf{z} = 1$. They are DPPs specified by the correlation kernels

$$K^{R_n}_{\mathbb{T}^2}(z, z') := \frac{1}{(2\pi)^2|\tau|} \sum_{j=1}^{n} \Psi^{R_n}_j(z) \overline{\Psi^{R_n}_j(z')}, \quad z, z' \in \mathbb{T}^2, \tag{6.28}$$

with respect to the Lebesgue measure $dz := d\mathrm{Re}\, z\, d\mathrm{Im}\, z$ *and they are denoted by* $(\Xi_{\mathbb{T}^2}^{R_n}, K_{\mathbb{T}^2}^{R_n}, dz)$, $R_n = A_{n-1}, B_n, B_n^{\vee}, C_n, C_n^{\vee}, BC_n, D_n$.

See Exercise 6.1.

We define the *reflection* and *shift* of DPP on $\mathbb{T}^2$ as follows.

Definition 6.1 Consider a DPP $(\Xi, K, \lambda(dz))$ on $\mathbb{T}^2$, where we write $\Xi(\cdot) = \sum_j \delta_{Z_j}(\cdot)$.

(i) The *reflection operator* $\mathcal{R}$ is defined by

$$\mathcal{R}\Xi := \sum_j \delta_{-Z_j},$$

$$\mathcal{R}K(z, z') := K(-z, -z'),$$

$$\mathcal{R}\lambda(dz) := \lambda(-dz).$$

We write $(\mathcal{R}\Xi, \mathcal{R}K, \mathcal{R}\lambda(dz))$ simply as $\mathcal{R}(\Xi, K, \lambda(dz))$.

(ii) For $u \in \mathbb{C}$, the *shift operator* by u, $\mathcal{S}_u$ is defined by

$$\mathcal{S}_u \Xi := \sum_j \delta_{Z_j - u},$$

$$\mathcal{S}_u K(z, z') := K(z + u, z' + u),$$

$$\mathcal{S}_u \lambda(dz) := \lambda(u + dz).$$

We write $(\mathcal{S}_u \Xi, \mathcal{S}_u K, \mathcal{S}_u \lambda(dz))$ simply as $\mathcal{S}_u(\Xi, K, \lambda(dz))$.

By definition of the seven DPPs on $\mathbb{T}^2$ given by Theorem 6.1,

$$\mathcal{S}_{2\pi}(\Xi_{\mathbb{T}^2}^{R_n}, K_{\mathbb{T}^2}^{R_n}, dz) = \mathcal{S}_{2\pi|\tau|i}(\Xi_{\mathbb{T}^2}^{R_n}, K_{\mathbb{T}^2}^{R_n}, dz) = (\Xi_{\mathbb{T}^2}^{R_n}, K_{\mathbb{T}^2}^{R_n}, dz),$$

for $R_n = A_{n-1}, B_n, B_n^{\vee}, C_n, C_n^{\vee}, BC_n, D_n$. In addition, we can prove the following symmetry which characterizes the seven types of DPPs on $\mathbb{T}^2$.

Proposition 6.2 *(i) For* $R_n = B_n, B_n^{\vee}, C_n, C_n^{\vee}, BC_n, D_n$, *the reflection invariance is established:*

$$\mathcal{R}(\Xi_{\mathbb{T}^2}^{R_n}, K_{\mathbb{T}^2}^{R_n}, dz) \overset{(law)}{=} (\Xi_{\mathbb{T}^2}^{R_n}, K_{\mathbb{T}^2}^{R_n}, dz).$$

(ii) The following equalities showing shift invariance are satisfied:

$$\mathcal{S}_{2\pi/n}(\Xi_{\mathbb{T}^2}^{A_{n-1}}, K_{\mathbb{T}^2}^{A_{n-1}}, dz) \overset{(law)}{=} (\Xi_{\mathbb{T}^2}^{A_{n-1}}, K_{\mathbb{T}^2}^{A_{n-1}}, dz), \tag{6.29}$$

$$\mathcal{S}_{2\pi|\tau|i/n}(\Xi_{\mathbb{T}^2}^{A_{n-1}}, K_{\mathbb{T}^2}^{A_{n-1}}, dz) \overset{(law)}{=} (\Xi_{\mathbb{T}^2}^{A_{n-1}}, K_{\mathbb{T}^2}^{A_{n-1}}, dz), \tag{6.30}$$

$$\mathcal{S}_{\pi}(\Xi_{\mathbb{T}^2}^{R_n}, K_{\mathbb{T}^2}^{R_n}, dz) \overset{(law)}{=} (\Xi_{\mathbb{T}^2}^{R_n}, K_{\mathbb{T}^2}^{R_n}, dz), \quad R_n = B_n^{\vee}, C_n, D_n, \tag{6.31}$$

$$\mathcal{S}_{\pi|\tau|i}(\Xi_{\mathbb{T}^2}^{R_n}, K_{\mathbb{T}^2}^{R_n}, dz) \overset{(law)}{=} (\Xi_{\mathbb{T}^2}^{R_n}, K_{\mathbb{T}^2}^{R_n}, dz), \quad R_n = C_n, C_n^{\vee}, D_n. \tag{6.32}$$

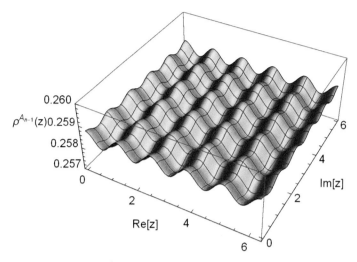

Fig. 6.2 The density of points $\rho_{\mathbb{T}^2}^{A_{n-1}}(z)$ is plotted on the fundamental domain $D_{(2\pi, 2\pi|\tau|)}$ in $\mathbb{C}$ for $n = 5$ and $|\tau| = 0.95$. As shown by Proposition 6.2 (ii), $n = 5$ oscillations are observed both in the real and the imaginary directions. Note that the amplitudes are very small. The amplitudes of oscillation in the imaginary (resp. real) direction becomes much smaller than that in the real (resp. imaginary) direction as t decreases from 1 (resp. increases from 1). There are no zeros

(iii) *The densities of points* $\rho_{\mathbb{T}^2}^{R_n}(z) = K_{\mathbb{T}^2}^{R_n}(z, z), z \in \mathbb{T}^2,$ *with respect to the Lebesgue measure dz have the following zeros:*

$$\rho_{\mathbb{T}^2}^{B_n}(0) = 0,$$
$$\rho_{\mathbb{T}^2}^{B_n^\vee}(0) = \rho_{\mathbb{T}^2}^{B_n^\vee}(\pi) = 0,$$
$$\rho_{\mathbb{T}^2}^{C_n^\vee}(0) = \rho_{\mathbb{T}^2}^{C_n^\vee}(\pi|\tau|i) = 0,$$
$$\rho_{\mathbb{T}^2}^{BC_n}(0) = \rho_{\mathbb{T}^2}^{BC_n}(\pi|\tau|i) = \rho_{\mathbb{T}^2}^{BC_n}(\pi + \pi|\tau|i) = 0,$$
$$\rho_{\mathbb{T}^2}^{C_n}(0) = \rho_{\mathbb{T}^2}^{C_n}(\pi) = \rho_{\mathbb{T}^2}^{C_n}(\pi|\tau|i) = \rho_{\mathbb{T}^2}^{C_n}(\pi + \pi|\tau|i) = 0.$$

By (6.26) of Lemma 6.5, Proposition 6.2 (i) is immediately proved. The equalities in law (6.29) and (6.31) of Proposition 6.2 (ii) are readily concluded from (3.39) of Exercise 3.1 and (3.43) of Exercise 3.3, respectively. The equalities in law (6.30) and (6.32) of Proposition 6.2 (ii) will be proved using (3.40), (3.41) of Exercise 3.1 and (3.44) of Exercise 3.3, respectively (Exercises 6.2 and 6.3). The facts (3.45)–(3.48) in Exercise 3.4 imply Proposition 6.2 (iii). The densities of points $\rho_{\mathbb{T}^2}^{R_n}(z)$ are plotted on the fundamental domain $D_{(2\pi, 2\pi|\tau|)}$ in Figs. 6.2–6.8.

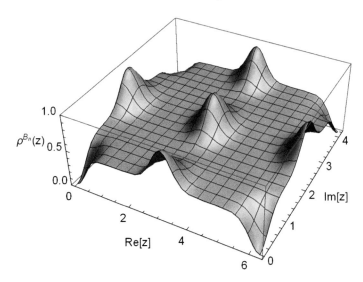

Fig. 6.3 The density of points $\rho_{\mathbb{T}^2}^{B_n}(z)$ is plotted on the fundamental domain $D_{(2\pi,2\pi|\tau|)}$ in $\mathbb{C}$ for $n = 10$ and $|\tau| = 2/3$. As mentioned in Proposition 6.2 (iii), there is a zero only at the origin $z = 0$ in the fundamental domain

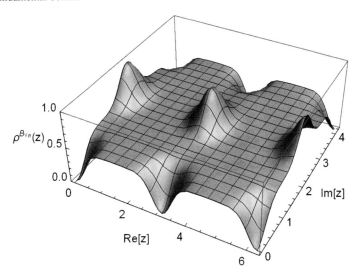

Fig. 6.4 The density of points $\rho_{\mathbb{T}^2}^{B_n^\vee}(z)$ is plotted on the fundamental domain $D_{(2\pi,2\pi|\tau|)}$ in $\mathbb{C}$ for $n = 10$ and $|\tau| = 2/3$. As shown by Proposition 6.2 (ii), the periodicity with period π is observed in the real direction. As mentioned in Proposition 6.2 (iii), there are two zeros at $z = 0$ and π in the fundamental domain

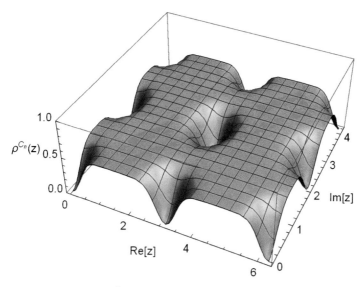

Fig. 6.5 The density of points $\rho_{\mathbb{T}^2}^{C_n}(z)$ is plotted on the fundamental domain $D_{(2\pi,2\pi|\tau|)}$ in $\mathbb{C}$ for $n = 10$ and $|\tau| = 2/3$. As shown by Proposition 6.2 (ii), the periodicity with period π is observed in the real direction as well as that with period $\pi|\tau|$ is observed in the imaginary direction. As mentioned in Proposition 6.2 (iii), there are four zeros at $z = 0, \pi, \pi|\tau|i$, and $\pi + \pi|\tau|i$ in the fundamental domain

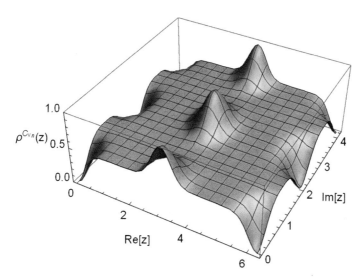

Fig. 6.6 The density of points $\rho_{\mathbb{T}^2}^{C_n^\vee}(z)$ is plotted on the fundamental domain $D_{(2\pi,2\pi|\tau|)}$ in $\mathbb{C}$ for $n = 10$ and $|\tau| = 2/3$. As shown by Proposition 6.2 (ii), the periodicity with period $\pi|\tau|$ is observed in the imaginary direction. As mentioned in Proposition 6.2 (iii), there are two zeros at $z = 0$ and $\pi|\tau|i$ in the fundamental domain

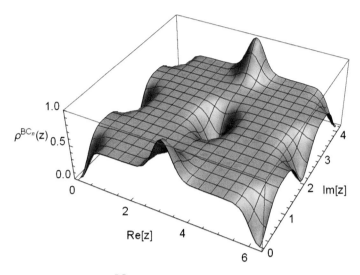

Fig. 6.7 The density of points $\rho_{\mathbb{T}^2}^{BC_n}(z)$ is plotted on the fundamental domain $D_{(2\pi, 2\pi|\tau|)}$ in $\mathbb{C}$ for $n = 10$ and $|\tau| = 2/3$. As mentioned in Proposition 6.2 (iii), there are three zeros at $z = 0, \pi|\tau|i$ and $\pi + \pi|\tau|i$ in the fundamental domain

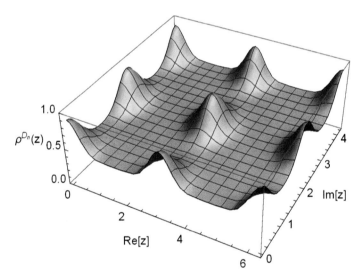

Fig. 6.8 The density of points $\rho_{\mathbb{T}^2}^{D_n}(z)$ is plotted on the fundamental domain $D_{(2\pi, 2\pi|\tau|)}$ in $\mathbb{C}$ for $n = 10$ and $|\tau| = 2/3$. As shown by Proposition 6.2 (ii), the periodicity with period π is observed in the real direction, and that with period $\pi|\tau|$ is observed in the imaginary direction. There are no zeros

6.3 Three Types of Ginibre DPPs

We note that the periods $2\pi/n \in (0, \infty)$ and $2\pi|\tau|i/n \in i(0, \infty)$ of $(\Xi_{\mathbb{T}^2}^{A_{n-1}}, K_{\mathbb{T}^2}^{A_{n-1}}, dz)$ shown by Proposition 6.2 (ii) become zeros as $n \to \infty$. Hence, at the $n \to \infty$ limit of $(\Xi_{\mathbb{T}^2}^{A_{n-1}}, K_{\mathbb{T}^2}^{A_{n-1}}, dz)$, we expect to obtain a uniform system of an infinite number of points on $\mathbb{C}$.

Here we introduce three kinds of infinite DPPs on $\mathbb{C}$. Let the reference measure be the *complex standard normal distribution* with mean zero, $\lambda_N(dz) := \dfrac{1}{\pi}e^{-|z|^2}dz$, and put

$$\mathcal{K}_{\text{Ginibre}}^A(z, z') := e^{z\bar{z'}},$$

$$\mathcal{K}_{\text{Ginibre}}^C(z, z') := \sinh(z\bar{z'}) = \frac{1}{2}(e^{z\bar{z'}} - e^{-z\bar{z'}}),$$

$$\mathcal{K}_{\text{Ginibre}}^D(z, z') := \cosh(z\bar{z'}) = \frac{1}{2}(e^{z\bar{z'}} + e^{-z\bar{z'}}), \quad z, z' \in \mathbb{C}. \tag{6.33}$$

Then the *Ginibre DPPs of type R* are defined by $(\Xi_{\text{Ginibre}}^R, \mathcal{K}_{\text{Ginibre}}^R, \lambda_N(dz))$ for $R = A, C$, and D, respectively. The Ginibre DPP of type A describes the eigenvalue distribution of the Gaussian complex random matrix in the bulk scaling limit [23]. The density of points is uniform with respect to the Lebesgue measure dz on $\mathbb{C}$ as

$$\rho_{\text{Ginibre}}^A(z)dz = \mathcal{K}_{\text{Ginibre}}^A(z, z)\lambda_N(dz) = \frac{1}{\pi}dz, \quad z \in \mathbb{C}.$$

On the other hand, the Ginibre DPPs of types C and D are rotationally symmetric around the origin, but non-uniform on $\mathbb{C}$. The densities of points with respect to the Lebesgue measure dz on $\mathbb{C}$ are given by

$$\rho_{\text{Ginibre}}^C(z)dz = \mathcal{K}_{\text{Ginibre}}^C(z, z)\lambda_N(dz) = \frac{1}{2\pi}(1 - e^{-2|z|^2})dz, \quad z \in \mathbb{C},$$

$$\rho_{\text{Ginibre}}^D(z)dz = \mathcal{K}_{\text{Ginibre}}^D(z, z)\lambda_N(dz) = \frac{1}{2\pi}(1 + e^{-2|z|^2})dz, \quad z \in \mathbb{C}. \tag{6.34}$$

It is obvious that $\rho_{\text{Ginibre}}^C(0) = 0$ and $\rho_{\text{Ginibre}}^C(z)$ is a strictly increasing function of $|z|$ with $\lim_{|z|\to\infty} \rho_{\text{Ginibre}}^C(z) = 1/2\pi$, and that $\rho_{\text{Ginibre}}^D(0) = 1/\pi$ and $\rho_{\text{Ginibre}}^D(z)$ is a strictly decreasing function of $|z|$ with $\lim_{|z|\to\infty} \rho_{\text{Ginibre}}^D(z) = 1/2\pi$. They were first reported in [39].

We can prove the following limit theorems.

Proposition 6.3 *The following weak convergence is established:*

$$\frac{1}{2}\sqrt{\frac{n}{\pi|\tau|}} \circ \left(\Xi_{\mathbb{T}^2}^{A_{n-1}}, K_{\mathbb{T}^2}^{A_{n-1}}, dz\right) \overset{n\to\infty}{\Longrightarrow} \left(\Xi_{\text{Ginibre}}^A, \mathcal{K}_{\text{Ginibre}}^A, \lambda_N(dz)\right), \tag{6.35}$$

$$\sqrt{\frac{n}{2\pi|\tau|}} \circ \left(\Xi_{\mathbb{T}^2}^{R_n}, \mathcal{K}_{\mathbb{T}^2}^{R_n}, dz\right) \overset{n\to\infty}{\Longrightarrow} \left(\Xi_{\text{Ginibre}}^{C}, \mathcal{K}_{\text{Ginibre}}^{C}, \lambda_N(dz)\right),$$

$$R_n = B_n, B_n^\vee, C_n, C_n^\vee, BC_n, \tag{6.36}$$

$$\sqrt{\frac{n}{2\pi|\tau|}} \circ \left(\Xi_{\mathbb{T}^2}^{D_n}, \mathcal{K}_{\mathbb{T}^2}^{D_n}, dx\right) \overset{n\to\infty}{\Longrightarrow} \left(\Xi_{\text{Ginibre}}^{D}, \mathcal{K}_{\text{Ginibre}}^{D}, \lambda_N(dz)\right). \tag{6.37}$$

Proof First we prove (6.35). We notice that

$$\begin{aligned}
\psi_j^{A_{n-1}}(e^{iz}; p, (-1)^n p^{n/2}) &= e^{i(j-1)z}\theta(-p^{j-1+n/2}e^{inz}; p^n) \\
&= e^{i(j-1)z}\theta(p^n(-p^{j-1-n/2}e^{inz}); p^n) \\
&= e^{\pi|\tau|(j-1-n/2)}e^{i(j-1-n/2)z} \\
&\quad \times e^{\pi|\tau|(j-1-n/2)}e^{-inz/2}\theta(-e^{-2\pi|\tau|(j-1-n/2)}e^{inz}; e^{-2n\pi|\tau|}),
\end{aligned}$$

$j = 1, \ldots, n$, where (1.10) and (6.7) are used. We use the equality

$$-\pi|\tau|(j-1)^2/n + \pi|\tau|(j-1-n/2) = -\pi|\tau|\{(j-1-n/2)/\sqrt{n}\}^2 - n\pi|\tau|/4.$$

Since $C_1(n|\tau|) \sim (2|\tau|)^{-1/2}n^{-1/2}$ as $n \to \infty$ by the definition (6.13), we have the estimate

$$\Psi_j^{A_{n-1}}\left(2\sqrt{\frac{\pi|\tau|}{n}}z\right) \sim e^{-n\pi|\tau|/4}(2|\tau|)^{1/4}n^{1/4}e^{-y^2}$$

$$\times \exp\left[-\pi|\tau|\left(\frac{j-1-n/2}{\sqrt{n}}\right)^2 + 2i\sqrt{\pi|\tau|}z\frac{j-1-n/2}{\sqrt{n}}\right]$$

$$\times \left\{e^{-i\sqrt{n\pi|\tau|}z}\exp\left(\pi|\tau|\sqrt{n}\frac{j-1-n/2}{\sqrt{n}}\right) + e^{i\sqrt{n\pi|\tau|}z}\exp\left(-\pi|\tau|\sqrt{n}\frac{j-1-n/2}{\sqrt{n}}\right)\right\}$$

as $n \to \infty$. Hence we can see that

$$K_{\mathbb{T}^2}^{A_{n-1}}\left(2\sqrt{\frac{\pi|\tau|}{n}}z, 2\sqrt{\frac{\pi|\tau|}{n}}z'\right)\left(2\sqrt{\frac{\pi|\tau|}{n}}\right)^2$$

$$\sim e^{-n\pi|\tau|/2}(2|\tau|)^{1/2}e^{-(y^2+y'^2)}4\pi|\tau|\frac{1}{(2\pi)^2|\tau|}$$

$$\times \frac{1}{\sqrt{n}}\sum_{j=1}^{n}\exp\left[-2\pi|\tau|\left(\frac{j-1-n/2}{\sqrt{n}}\right)^2 + 2i\sqrt{\pi|\tau|}\frac{j-1-n/2}{\sqrt{n}}(z-\overline{z'})\right]$$

$$\times \left[e^{-i\sqrt{n\pi|\tau|}z}\exp\left(\pi|\tau|\sqrt{n}\frac{j-1-n/2}{\sqrt{n}}\right) + e^{i\sqrt{n\pi|\tau|}z}\exp\left(-\pi|\tau|\sqrt{n}\frac{j-1-n/2}{\sqrt{n}}\right)\right]$$

$$\times \left[e^{i\sqrt{n\pi|\tau|}\overline{z'}}\exp\left(\pi|\tau|\sqrt{n}\frac{j-1-n/2}{\sqrt{n}}\right) + e^{-i\sqrt{n\pi|\tau|}\overline{z'}}\exp\left(-\pi|\tau|\sqrt{n}\frac{j-1-n/2}{\sqrt{n}}\right)\right]$$

$$\sim e^{-n\pi|\tau|/2}\frac{(2|\tau|)^{1/2}}{\pi}e^{-(y^2+y'^2)}$$

$$\times \left\{ e^{-i\sqrt{n\pi|\tau|}(z-\bar{z}')} I_+ + 2\cos(\sqrt{n\pi|\tau|}(z+\bar{z}'))I_0 + e^{i\sqrt{n\pi|\tau|}(z-\bar{z}')} I_- \right\},$$

as $n \to \infty$. Here

$$I_\pm := \int_{-\sqrt{n}/2}^{\sqrt{n}/2} e^{-2\pi|\tau|u^2 + 2i\sqrt{\pi|\tau|}(z-\bar{z}')u \pm 2\pi|\tau|\sqrt{n}u}\, du$$

$$= e^{n\pi|\tau|/2} e^{\pm i\sqrt{n\pi|\tau|}(z-\bar{z}')} e^{-(z-\bar{z}')^2/2} \int_{a_\pm - i(z-\bar{z}')/2\sqrt{\pi|\tau|}}^{b_\pm - i(z-\bar{z}')/2\sqrt{\pi|\tau|}} e^{-2\pi|\tau|v^2}\, dv$$

with $a_+ = -b_- = -\sqrt{n}$, $b_+ = a_- = 0$, and

$$I_0 := \int_{-\sqrt{n}/2}^{\sqrt{n}/2} e^{-2\pi|\tau|u^2 + 2i\sqrt{\pi|\tau|}(z-\bar{z}')u}\, du$$

$$= e^{-(z-\bar{z}')^2/2} \int_{-\sqrt{n}/2 - i(z-\bar{z}')/2\sqrt{\pi|\tau|}}^{\sqrt{n}/2 - i(z-\bar{z}')/2\sqrt{\pi|\tau|}} e^{-2\pi|\tau|v^2}\, dv.$$

We see

$$I_\pm \sim e^{n\pi|\tau|/2} \frac{1}{2} \frac{1}{(2|\tau|)^{1/2}} e^{\pm i\sqrt{n\pi|\tau|}(z-\bar{z}')} e^{-(z-\bar{z}')^2/2},$$

$$I_0 \to \frac{1}{(2|\tau|)^{1/2}} e^{-(z-\bar{z}')^2/2}, \quad \text{as } n \to \infty.$$

Therefore, we have

$$K_{\mathbb{T}^2}^{A_{n-1}}\left(2\sqrt{\frac{\pi|\tau|}{n}}z, 2\sqrt{\frac{\pi|\tau|}{n}}z'\right)\left(2\sqrt{\frac{\pi|\tau|}{n}}\right)^2 \to \frac{1}{\pi} e^{-(y^2+y'^2)-(z-\bar{z}')^2/2}.$$

The convergence is uniform in any compact domain in $\mathbb{C}$. Since

$$\frac{1}{\pi} e^{-(y^2+y'^2)-(z-\bar{z}')^2/2} = \frac{e^{-ixy}}{e^{-ix'y'}} e^{z\bar{z}'} \frac{1}{\pi} e^{-(|z|^2+|z'|^2)/2},$$

the gauge invariance of DPP (Lemma 5.3) and the equivalence (5.31) between DPPs implies (6.35).

Next we prove (6.36) and (6.37). For $R_n = B_n$, $B_n^\vee$, C_n, $C_n^\vee$, BC_n, and D_n, we see that

$$\Psi_j^{R_n}\left(\sqrt{\frac{2\pi|\tau|}{n}}\,z\right) \sim \frac{1}{\sqrt{c_j}}|\tau|^{1/4}n^{1/4}e^{-y^2}e^{-\pi|\tau|(j-n)^2/2n}$$

$$\times\left[e^{i(j-n)\sqrt{2\pi|\tau|/nz}}\left(1+\sigma_1 e^{-2\pi|\tau|j}e^{i2n\sqrt{2\pi|\tau|/nz}}\right)\right.$$

$$\left.+\sigma_2 e^{-i(j-n)\sqrt{2\pi|\tau|/nz}}\left(1+\sigma_1 e^{-2\pi|\tau|j}e^{-i2n\sqrt{2\pi|\tau|/nz}}\right)\right]$$

$$=\frac{1}{\sqrt{c_j}}|\tau|^{1/4}n^{1/4}e^{-y^2}e^{-n\pi|\tau|/2}e^{-\pi|\tau|j^2/2n}$$

$$\times\left[e^{-in\sqrt{2\pi|\tau|/nz}}e^{i\sqrt{2\pi|\tau|/nz}j}e^{\pi|\tau|j}+\sigma_1 e^{-in\sqrt{2\pi|\tau|/nz}}e^{i\sqrt{2\pi|\tau|/nz}j}e^{-\pi|\tau|j}\right.$$

$$\left.+\sigma_2 e^{in\sqrt{2\pi|\tau|/nz}}e^{-i\sqrt{2\pi|\tau|/nz}j}e^{\pi|\tau|j}+\sigma_1\sigma_2 e^{-in\sqrt{2\pi|\tau|/nz}}e^{-i\sqrt{2\pi|\tau|/nz}j}e^{-\pi|\tau|j}\right],$$

as $n\to\infty$. Hence we have

$$K_{\mathbb{T}^2}^{R_n}\left(\sqrt{\frac{2\pi|\tau|}{n}}\,z,\sqrt{\frac{2\pi|\tau|}{n}}\,z'\right)\left(\sqrt{\frac{2\pi|\tau|}{n}}\right)$$

$$\sim\frac{|\tau|^{1/2}}{2\pi}e^{-(y^2+y'^2)}\sum_{s_1=\pm1}\sum_{s_2=\pm1}\left\{(S_{s_1,s_2,-1}^{(1)}+\sigma_2 S_{s_1,s_2,1}^{(1)})+\sigma_1(S_{s_1,s_2,-1}^{(2)}+\sigma_2 S_{s_1,s_2,1}^{(2)})\right\},$$

as $n\to\infty$, where

$$S_{s_1,s_2,s_3}^{(1)}=e^{-n\pi|\tau|}e^{-s_1 s_2 i\sqrt{2\pi n|\tau|}(z+s_3\overline{z'})}$$

$$\times\frac{1}{\sqrt{n}}\sum_{j=1}^{n}\frac{1}{c_j}\exp\left[-\pi|\tau|\left(\frac{j}{\sqrt{n}}\right)^2+\left\{s_1 2\pi|\tau|\sqrt{n}s_2 i\sqrt{2\pi|\tau|}(z+s_3\overline{z'})\right\}\frac{j}{\sqrt{n}}\right],$$

$$S_{s_1,s_2,s_3}^{(2)}=e^{-n\pi|\tau|}e^{s_1 i\sqrt{2\pi n|\tau|}(z-s_3\overline{z'})}$$

$$\times\frac{1}{\sqrt{n}}\sum_{j=1}^{n}\frac{1}{c_j}\exp\left[-\pi|\tau|\left(\frac{j}{\sqrt{n}}\right)^2+s_2 i\sqrt{2\pi|\tau|}(z-s_3\overline{z'})\right\}\frac{j}{\sqrt{n}}\right].$$

We can see that

$$S_{s_1,s_2,s_3}^{(1)}\sim e^{-n\pi|\tau|}e^{-s_1 s_2 i\sqrt{2\pi n|\tau|}(z+s_3\overline{z'})}$$

$$\times\int_0^{\sqrt{n}}e^{-\pi|\tau|w^2+\{s_1 2\pi|\tau|\sqrt{n}+s_2 i\sqrt{2\pi|\tau|}(z+s_3\overline{z'})\}w}dw$$

$$=e^{(z+s_3\overline{z'})^2/2}J_{s_1,s_2,s_3}^{(1)},\quad\text{as }n\to\infty,$$

where

$$J_{s_1,s_2,s_3}^{(1)}=\int_{-(s_1\sqrt{n}+s_2 i(z+s_3\overline{z'})/\sqrt{2\pi|\tau|}}^{\sqrt{n}-(s_1\sqrt{n}+s_2 i(z+s_3\overline{z'})/\sqrt{2\pi|\tau|}}e^{-\pi|\tau|u^2}du.$$

It is easy to verify that as $n \to \infty$,

$$J^{(1)}_{1,1,s_3} + J^{(1)}_{1,-1,s_3} = \int_{-\sqrt{n}-i(z+s_3\overline{z'})/\sqrt{2\pi|\tau|}}^{\sqrt{n}-i(z+s_3\overline{z'})/\sqrt{2\pi|\tau|}} e^{-\pi|\tau|u^2} du \to \int_{-\infty}^{\infty} e^{-\pi|\tau|u^2} du = |\tau|^{1/2},$$

and that

$$J^{(1)}_{-1,s_2,s_3} = \int_{2\sqrt{n}-s_2i(z+s_3\overline{z'})/\sqrt{2\pi|\tau|}}^{\sqrt{n}-s_2i(z+s_3\overline{z'})/\sqrt{2\pi|\tau|}} e^{-\pi|\tau|u^2} du \to 0.$$

On the other hand, due to the factor $e^{-n\pi|\tau|}$,

$$S^{(2)}_{s_1,s_2,s_3} \sim e^{-n\pi|\tau|} e^{s_1 i \sqrt{2\pi n|\tau|}(z+s_3\overline{z'})} \int_0^{\sqrt{n}} e^{-\pi|\tau|w^2 + s_2 i \sqrt{2\pi|\tau|}(z-s_3\overline{z'})w} dw \to 0,$$

as $n \to \infty$. Therefore, we obtain for $R_n = B_n, B_n^\vee, C_n, C_n^\vee, BC_n$, and D_n,

$$K^{R_n}_{\mathbb{T}^2}\left(\sqrt{\frac{2\pi|\tau|}{n}}z, \sqrt{\frac{2\pi|\tau|}{n}}z'\right)\left(\sqrt{\frac{2\pi|\tau|}{n}}\right) \to \frac{1}{2\pi} e^{-(y^2+y'^2)}\left\{e^{-(z-\overline{z'})^2} + \sigma_2 e^{-(z+\overline{z'})^2}\right\},$$

as $n \to \infty$. The convergence is uniform in any compact domain in $\mathbb{C}$. We can readily confirm that

$$\frac{1}{2\pi} e^{-(y^2+y'^2)}\left\{e^{-(z-\overline{z'})^2} + \sigma_2 e^{-(z+\overline{z'})^2}\right\} = \frac{e^{-ixy}}{e^{-ix'y'}} \frac{1}{2}(e^{z\overline{z'}} + \sigma_2 e^{-z\overline{z'}}) e^{-(|z|^2+|z'|^2)/2},$$

and

$$\frac{1}{2}(e^{z\overline{z'}} + \sigma_2 e^{-z\overline{z'}}) = \begin{cases} \sinh(z\overline{z'}) = K^C_{\text{Ginibre}}(z, z'), & \text{if } \sigma_2 = -1, \\ \cosh(z\overline{z'}) = K^D_{\text{Ginibre}}(z, z'), & \text{if } \sigma_2 = 1. \end{cases}$$

Since $\sigma_2^{R_n} = -1$ for $R_n = B_n, B_n^\vee, C_n, C_n^\vee, BC_n$ and $\sigma_2^{D_n} = 1$ as (3.12), (6.36) and (6.37) are proved. $\qquad\square$

Exercises

6.1 For $\Psi_j^{A_{n-1}}(z)$ defined by (6.18), prove the equalities

$$\Psi_{j\pm n}^{A_{n-1}}(z) = \Psi_j^{A_{n-1}}(z), \quad j = 1, \dots, n. \tag{6.38}$$

In other words,

$$\Psi_j^{A_{n-1}}(z) = \Psi_k^{A_{n-1}}(z) \quad \text{if } j = k \mod n.$$

This implies that in the formula (6.28) for $K_{\mathbb{T}^2}^{A_{n-1}}$ in Theorem 6.1, the range of j in the summation, which is given by $\{1, \ldots, n\}$ there, can be replaced by any n integers which are different from each other modulo n.

6.2 Prove (6.30) of Proposition 6.2 (ii) using (3.40) and (3.41) of Exercise 3.1.

6.3 Prove (6.32) of Proposition 6.2 (ii) using (3.44) of Exercise 3.3.

6.4 As noted in Exercise 6.1 above, $K_{\mathbb{T}^2}^{A_n}$ defined by (6.28) for $R_n = A_{n-1}$ is written as

$$K_{\mathbb{T}^2}^{A_{n-1}}(z, z') = \frac{1}{(2\pi)^2 |\tau|} \sum_{j=-[n/2]+1}^{[(n+1)/2]} \Psi_j^{A_{n-1}}(z) \overline{\Psi_j^{A_{n-1}}(z')}, \quad z, z' \in \mathbb{T}^2,$$

where $[x]$ denotes the largest integer less than or equal to x. Let

$$\varrho := \frac{n}{(2\pi)^2 |\tau|} = \text{constant},$$

and take the double limit $n \to \infty$ and $|\tau| \to \infty$. Show that in this scaling limit, we obtain a DPP with an infinite number of points on the surface of the cylinder

$$\mathbb{T} \times \mathbb{R} := \{(x, y) : x \in \mathbb{T}, y \in \mathbb{R}\}.$$

Choose an appropriate reference measure on $\mathbb{T} \times \mathbb{R}$ and determine the correlation kernel of the DPP obtained in this limit. (For other R_n, see [39] and [43, Sect. 4.4].)

Chapter 7
Future Problems

Abstract So far we have studied two kinds of random systems whose probability laws are expressed using the theta function. One of them is a family of one-dimensional stochastic processes consisting of seven types of noncolliding Brownian bridges. Another one is a family of two-dimensional point processes consisting of seven types of DPPs on $\mathbb{T}^2$. In this last chapter, we will address future problems concerning these two families of random systems. For the former systems, the key aspect of future direction will be dynamical property of DPPs. For the latter systems, it will be the connections to other statistical systems such as the one-component plasma models and the Gaussian free fields.

7.1 Stochastic Differential Equations for Dynamically Determinantal Processes

In Chap. 4, we defined the seven types of noncolliding Brownian bridges $(\mathbf{X}^{R_n}(t))_{t \in [0,T]}$ with time duration $T > 0$ for $R_n = A_{n-1}$, B_n, $B_n^{\vee}$, C_n, $C_n^{\vee}$, BC_n, D_n, whose single time probability densities are denoted by $\mathbf{p}_T^{R_n}(t, \mathbf{x})$;

$$\mathbf{P}_T^{R_n}(\mathbf{X}^{R_n}(t) \in d\mathbf{x}) = \mathbf{p}_T^{R_n}(t, \mathbf{x})d\mathbf{x}, \quad t \in [0, T].$$

They are generalizations of the one-dimensional Brownian bridge in the two directions: (i) the multi-variate generalization for n-particle systems, $n \in \mathbb{N}$, and (ii) the elliptic extensions such that $\mathbf{p}_T^{R_n}(t, \mathbf{x})$ are expressed using determinants of $n \times n$ matrices whose entries are given by the biorthogonal R_n theta functions of Rosegren and Schlosser (Propositions 4.2 and 4.3).

Here we recall the original Brownian bridge on $\mathbb{R}$ starting from u at time $t = 0$ and arriving at the same point u at the final time $T > 0$. The single time probability density is given by

$$p_T(t, x) = \frac{1}{Z}p(0, u; t, x)p(t, x; T, u), \quad t \in [0, T], \tag{7.1}$$

where p is the transition probability density of the one-dimensional standard Brownian motion (2.1) and $Z = p(0, u; T, u)$. We write this one-dimensional Brownian bridge as $(X(t))_{t \in [0,T]}$, where the initial and final values are pinned at u; $X(0) = X(T) = u$. One can show that this stochastic process satisfies the following *stochastic differential equation* (SDE):

$$dX(t) = dB(t) + \frac{u - X(t)}{T - t} dt, \quad t \in [0, T), \tag{7.2}$$

where $(B(t))_{t \geq 0}$ is the one-dimensional standard Brownian motion. (See, for instance, [3, Sect. IV.4, Part I].) We notice that the coefficient of the drift term $(u - x)/(T - t)|_{x = X(t)}$ in (7.2) is obtained by the *logarithmic derivative* of the last factor in (7.1),

$$\frac{\partial}{\partial x} \log p(t, x; T, u) = \frac{\partial}{\partial x} \log \frac{e^{-(u-x)^2/\{2(T-t)\}}}{\sqrt{2\pi(T - t)}}$$

$$= \frac{u - x}{T - t}.$$

For $\mathbf{p}_T^{A_{n-1}}(t, \mathbf{x})$ given by (4.21) in Proposition 4.2 and $\mathbf{p}_T^{R_n}(t, \mathbf{x})$ by (4.38) in Proposition 4.3, $R_n = B_n, B_n^\vee, C_n, C_n^\vee, BC_n, D_n$, the corresponding logarithmic derivatives are

$$\mathrm{Re} \frac{\partial}{\partial x_j} \det_{1 \leq k, \ell \leq n} \left[\psi_k^{A_{n-1}}(e^{ix\ell}; p_{n(T-t)}, r_{T-t}) \right], \quad j = 1, \ldots, n,$$

and

$$\mathrm{Re} \frac{\partial}{\partial x_j} \det_{1 \leq k, \ell \leq n} \left[\psi_k^{R_n}(e^{ix\ell}; p_{\mathcal{N}^{R_n}(T-t)}) \right], \quad j = 1, \ldots, n,$$

respectively. Notice that since the biorthogonal R_n theta functions are complex functions, we should take the real parts of the logarithmic derivatives. By the determinantal identities of Rosengren and Schlosser (Propositions 3.1 and 3.2) for the Macdonald denominators (Definitions 3.2 and 3.4), the SDEs will be given by the following: for $t \in [0, T)$,

$$dX_j^{A_{n-1}}(t) = dB_j^A(t) + \mathrm{Re} \frac{\partial}{\partial x_j} \log \theta(r_{T-t} e^{i \sum_{\ell=1}^n x_\ell}; p_{n(T-t)}) \Big|_{\mathbf{x} = \mathbf{X}^{A_{n-1}}(t)} dt$$

$$+ \mathrm{Re} \sum_{1 \leq k \leq n, k \neq j} \frac{\partial}{\partial x_j} \log \theta(e^{i(x_j - x_k)}; p_{n(T-t)}) \Big|_{\mathbf{x} = \mathbf{X}^{A_{n-1}}(t)} dt, \quad j = 1, \ldots, n, \tag{7.3}$$

$$dX_j^{B_n}(t) = dB_j^B(t) + \mathrm{Re} \frac{\partial}{\partial x_j} \log \theta(e^{ix_j}; p_{\mathcal{N}^{B_n}(T-t)}) \Big|_{\mathbf{x} = \mathbf{X}^{B_n}(t)} dt$$

$$+ \mathrm{Re} \sum_{1 \leq k \leq n, k \neq j} \left\{ \frac{\partial}{\partial x_j} \log \theta(e^{i(x_j - x_k)}; p_{\mathcal{N}^{B_n}(T-t)}) \right.$$

$$+ \frac{\partial}{\partial x_j} \log \theta (e^{i(x_j+x_k)}; p_{\mathcal{N}^{B_n}(T-t)}) \Big\} \Big|_{\mathbf{x}=\mathbf{X}^{B_n}(t)} dt, \quad j = 1, \ldots, n, \qquad (7.4)$$

$$dX_j^{B_n^\vee}(t) = dB_j^{B^\vee}(t) + \mathrm{Re}\, \frac{\partial}{\partial x_j} \log \theta (e^{2ix_j}; p_{2\mathcal{N}^{B_n^\vee}(T-t)}) \Big|_{\mathbf{x}=\mathbf{X}^{B_n^\vee}(t)} dt$$

$$+ \mathrm{Re} \sum_{1 \le k \le n, k \ne j} \Big\{ \frac{\partial}{\partial x_j} \log \theta (e^{i(x_j-x_k)}; p_{\mathcal{N}^{B_n^\vee}(T-t)})$$

$$+ \frac{\partial}{\partial x_j} \log \theta (e^{i(x_j+x_k)}; p_{\mathcal{N}^{B_n^\vee}(T-t)}) \Big\} \Big|_{\mathbf{x}=\mathbf{X}^{B_n^\vee}(t)} dt, \quad j = 1, \ldots, n, \qquad (7.5)$$

$$dX_j^{C_n}(t) = dB_j^{C}(t) + \mathrm{Re}\, \frac{\partial}{\partial x_j} \log \theta (e^{2ix_j}; p_{\mathcal{N}^{C_n}(T-t)}) \Big|_{\mathbf{x}=\mathbf{X}^{C_n}(t)} dt$$

$$+ \mathrm{Re} \sum_{1 \le k \le n, k \ne j} \Big\{ \frac{\partial}{\partial x_j} \log \theta (e^{i(x_j-x_k)}; p_{\mathcal{N}^{C_n}(T-t)})$$

$$+ \frac{\partial}{\partial x_j} \log \theta (e^{i(x_j+x_k)}; p_{\mathcal{N}^{C_n}(T-t)}) \Big\} \Big|_{\mathbf{x}=\mathbf{X}^{C_n}(t)} dt, \quad j = 1, \ldots, n, \qquad (7.6)$$

$$dX_j^{C_n^\vee}(t) = dB_j^{C^\vee}(t) + \mathrm{Re}\, \frac{\partial}{\partial x_j} \log \theta (e^{ix_j}; p_{\mathcal{N}^{C_n^\vee}(T-t)/2}) \Big|_{\mathbf{x}=\mathbf{X}^{C_n^\vee}(t)} dt$$

$$+ \mathrm{Re} \sum_{1 \le k \le n, k \ne j} \Big\{ \frac{\partial}{\partial x_j} \log \theta (e^{i(x_j-x_k)}; p_{\mathcal{N}^{C_n^\vee}(T-t)})$$

$$+ \frac{\partial}{\partial x_j} \log \theta (e^{i(x_j+x_k)}; p_{\mathcal{N}^{C_n^\vee}(T-t)}) \Big\} \Big|_{\mathbf{x}=\mathbf{X}^{C_n^\vee}(t)} dt, \quad j = 1, \ldots, n, \qquad (7.7)$$

$$dX_j^{BC_n}(t) = dB_j^{BC}(t) + \mathrm{Re}\, \frac{\partial}{\partial x_j} \log \theta (e^{ix_j}; p_{\mathcal{N}^{BC_n}(T-t)}) \Big|_{\mathbf{x}=\mathbf{X}^{BC_n}(t)} dt$$

$$+ \mathrm{Re}\, \frac{\partial}{\partial x_j} \log \theta (p_{\mathcal{N}^{BC_n}(T-t)} e^{2ix_j}; p_{2\mathcal{N}^{BC_n}(T-t)}) \Big|_{\mathbf{x}=\mathbf{X}^{BC_n}(t)} dt$$

$$+ \mathrm{Re} \sum_{1 \le k \le n, k \ne j} \Big\{ \frac{\partial}{\partial x_j} \log \theta (e^{i(x_j-x_k)}; p_{\mathcal{N}^{BC_n}(T-t)})$$

$$+ \frac{\partial}{\partial x_j} \log \theta (e^{i(x_j+x_k)}; p_{\mathcal{N}^{BC_n}(T-t)}) \Big\} \Big|_{\mathbf{x}=\mathbf{X}^{BC_n}(t)} dt, \quad j = 1, \ldots, n, \qquad (7.8)$$

$$dX_j^{D_n}(t) = dB_j^{D}(t) + \mathrm{Re} \sum_{1 \le k \le n, k \ne j} \Big\{ \frac{\partial}{\partial x_j} \log \theta (e^{i(x_j-x_k)}; p_{\mathcal{N}^{D_n}(T-t)})$$

$$+ \frac{\partial}{\partial x_j} \log \theta (e^{i(x_j+x_k)}; p_{\mathcal{N}^{D_n}(T-t)}) \Big\} \Big|_{\mathbf{x}=\mathbf{X}^{D_n}(t)} dt, \quad j = 1, \ldots, n, \qquad (7.9)$$

where $(B_j^R(t))_{t \ge 0}$, $j = 1, \ldots, n$, $R = A, B, B^\vee, C, C^\vee, BC, D$, are mutually independent one-dimensional standard Brownian motions, $\mathcal{N}^{R_n}$ and r_t are given by (3.9) and (4.19), and we impose a reflecting wall at 0 for $R_n = D_n$ and at π for $R_n = B_n$, $C_n^\vee$, BC_n, and D_n.

We can apply Jacobi's imaginary transformation (Lemma 2.1) to the theta functions in the drift terms. Then the asymptotic forms of the SDEs in $t \uparrow T$ are obtained as follows:

$$dX_j^{R_n}(t) \asymp dB_j^R(t) + \frac{u_j^{R_n} - X_j^{R_n}(t)}{T-t}, \quad j = 1, \ldots, n, \quad \text{as } t \uparrow T,$$

where $u_j^{A_{n-1}}$ and $u_j^{R_n}$, $j = 1, \ldots, n$, $R_n = B_n, B_n^\vee, C_n, C_n^\vee, BC_n, D_n$ are given by (4.3) and (4.24), respectively. The asymptotics prove that the final configurations of the present noncolliding Brownian bridges are indeed pinned at $\mathbf{X}^{R_n}(T) = \mathbf{u}^{R_n}$ as we have set in Chap. 4.

If we take the temporally homogeneous limit, $t \to \infty$, $T - t \to \infty$ in (7.3), $p_{n(T-t)} \to 0$ and $r_{T-t} \to 0$, and hence we see that

$$\frac{\partial}{\partial x_j} \log \theta(e^{i(x_j - x_k)}; p_{n(T-t)})$$

$$\to \frac{\partial}{\partial x_j} \log(1 - e^{i(x_j - x_k)}) = -\frac{i e^{i(x_j - x_k)}}{1 - e^{i(x_j - x_k)}} = \frac{1}{2} \cot\left(\frac{x_j - x_k}{2}\right) + \frac{i}{2}.$$

Hence (7.3) is reduced to

$$dY_j^{A_{n-1}}(t) = d\widetilde{B}_j^A(t) + \frac{1}{2} \sum_{1 \le k \le n, k \ne j} \cot\left\{\frac{1}{2}\left(Y_j^{A_{n-1}}(t) - Y_k^{A_{n-1}}(t)\right)\right\} dt, \quad j = 1, \ldots, n$$

$t \ge 0$. This system of SDEs is regarded as Dyson's Brownian motion model with parameter $\beta = 2$ on the one-dimensional torus $\mathbb{T}$ (i.e., the unit circle) [33]. Here *Dyson's Brownian motion model* with parameter β on $\mathbb{R}$, $\mathbf{Y}(t) = (Y_1(t), \ldots, Y_n(t))$, $t \ge 0$, is given by the solution of the following system of SDEs [36]:

$$dY_j(t) = dB_j(t) + \frac{\beta}{2} \sum_{1 \le k \le n, k \ne j} \frac{dt}{Y_j(t) - Y_k(t)}, \quad j = 1, \ldots, n, \quad t \ge 0.$$

In other words, the seven systems of SDEs (7.3)–(7.9) are elliptic extensions of Dyson's Brownian motion model with $\beta = 2$. Stochastic analysis of interacting particle systems (see, for instance, [24–26]) will be developed in the future to involve logarithmic derivatives of the theta functions in drift terms.

In Chap. 5 we have proved that, at each time $t \in [0, T]$, $\{\mathbf{X}^{R_n}(t)\}$ provide seven types of DPPs. The notion of DPP is extended dynamically in which spatio-temporal correlation functions are defined and expressed by determinants specified by a time-dependent correlation kernel [36, 46, 47]. By the previous studies [33–35, 37], it is expected that the seven types of noncolliding Brownian bridges are also *dynamically determinantal processes*. Explicit calculation of the spatio-temporal correlation kernels, study of dynamical properties of temporally inhomogeneous interacting particle systems, and construction of infinite-particle systems [47, 58–61, 76] will be fruitful future problems.

7.2 One-Component Plasma Models and Gaussian Free Fields

In Chap. 6, we studied seven kinds of DPPs on $\mathbb{T}^2$, whose probability densities are given by Theorem 6.1 in the form

$$\mathbf{p}_{\mathbb{T}^2}^{R_n}(\mathbf{z}) = \frac{1}{Z^{R_n}} \left| \det_{1 \le j,k \le n} [\Psi_j^{R_n}(z_k)] \right|^2, \quad \mathbf{z} = (z_1, \dots, z_n) \in (\mathbb{T}^2)^n. \tag{7.10}$$

By the definitions (6.18) and (6.19) of $\{\Psi_j^{R_n}\}_{j=1}^n$, we see that the above probability densities are proportional to the squares of determinants of the R_n theta functions, and hence also to the squares of the Macdonald denominators $|M^{R_n}|^2$ though Propositions 3.1 and 3.2. Hence, here we see $|M^{A_{n-1}}|^2$ and $|M^{D_n}|^2$ more carefully. (Notice that M^{D_n} can be regarded as the common factor of M^{R_n} for other five types, $R_n = B_n, B_n^\vee, C_n, C_n^\vee, BC_n$.) By Definitions 3.2 and 3.4, we have

$$|M^{A_{n-1}}(e^{i\mathbf{z}}; p)|^2 = \left| \prod_{1 \le j < k \le n} e^{iz_k} \theta(e^{i(z_j - z_k)}; p) \right|^2$$

$$= \prod_{\ell=1}^n e^{-2(\ell-1)y_\ell} \prod_{1 \le j < k \le n} |\theta(e^{i(z_j - z_k)}; p)|^2$$

$$= \exp\left[-2 \sum_{j=1}^n (j-1)y_j - 2 \sum_{1 \le j < k \le n} \Phi_0(z_j, z_k) \right],$$

and

$$|M^{D_n}(e^{i\mathbf{z}}; p)|^2 = \exp\left[2 \sum_{j=1}^n (n-j)y_j - 2 \sum_{1 \le j < k \le n} \Phi_0^\pm(z_j, z_k) \right],$$

where $y_j = \operatorname{Im} z_j$, $j = 1, \dots, n$ and

$$\Phi_0(z, z') = \Phi_0(z, z'; p) := -\log|\theta(e^{i(z-z')}; p)|,$$
$$\Phi_0^\pm(z, z') = \Phi_0^\pm(z, z'; p) := -\log|\theta(e^{i(z-z')}; p)| - \log|\theta(e^{i(z+z')}; p)|.$$

As $|z - z'| \to 0$, $\theta(e^{i(z-z')}; p) \sim \theta'(1; p)i(z - z') = -i(p; p)_\infty^2(z - z')$, where (1.14) was used. Hence, we have the asymptotics,

$$\Phi_0(z, z') \sim -\log|z - z'| - 2\log(p; p)_\infty,$$
$$\Phi_0^\pm(z, z') \sim -\log|z - z'| - 2\log(p; p)_\infty$$
$$\qquad - \frac{1}{2}\left\{ \log|\theta(e^{2iz}; p)| + \log|\theta(e^{2iz'}; p)| \right\} \quad \text{as } |z - z'| \to 0.$$

Therefore, if we define

$$\Phi(z, z') := \Phi_0(z, z') + 2\log(p; p)_\infty,$$

$$\Phi^\pm(z, z') := \Phi_0^\pm(z, z') + 2\log(p; p)_\infty + \frac{1}{2}\left\{\log|\theta(e^{2iz}; p)| + \log|\theta(e^{2iz'}; p)|\right\},$$

$$\tag{7.11}$$

then they show the common short-distance behavior

$$\Phi(z, z') \sim \Phi^\pm(z, z') \sim -\log|z - z'| \quad \text{as } |z - z'| \to 0,$$

and solve the *two-dimensional Poisson equation*

$$\Delta\Phi(z, z') = \Delta\Phi^\pm(z, z') = -2\pi\delta(z - z'), \quad z, z' \in \mathbb{C},$$

where $\Delta = 4\partial^2/\partial z\partial\bar{z} = \partial^2/\partial x^2 + \partial^2/\partial y^2$ for $z = x + iy, x, y \in \mathbb{R}$.

We regard Φ and $\Phi^\pm$ as *two-dimensional Coulomb potentials* and consider two types of n-particle systems in which each particle has a positive unit charge $+1$. We assume that the charged particles are mobile and interacting via the potential Φ or $\Phi^\pm$, but they are confined on the two-dimensional surface $\mathbb{T}^2$. The total energies of the particle–particle interaction in particle configuration $\mathbf{z} \in (\mathbb{T}^2)^n$ are given by

$$E_{pp}(\mathbf{z}) = \sum_{1 \le j < k \le n} \Phi(z_j, z_k)$$

$$= -\log\left(\prod_{1 \le j < k \le n} |\theta(e^{i(z_j - z_k)}; p)|\right) + n(n-1)\log(p; p)_\infty,$$

and

$$E_{pp}^\pm(\mathbf{z}) = \sum_{1 \le j < k \le n} \Phi^\pm(z_j, z_k)$$

$$= -\log\left(\prod_{1 \le j < k \le n} |\theta(e^{i(z_j - z_k)}; p)|\right) - \log\left(\prod_{1 \le j < k \le n} |\theta(e^{i(z_j + z_k)}; p)|\right)$$

$$+ n(n-1)\log(p; p)_\infty + \frac{n-1}{2}\log\left(\prod_{j=1}^n |\theta(e^{2iz_j}; p)|\right),$$

respectively.

We think that in both systems there are *uniform backgrounds* which are negatively charged. We assume that each background is consisting of n^- negative charges. Hence charge density of the background is given by $-n^-/4\pi^2|\tau|$ respect to the Lebesgue measure dz on $\mathbb{T}^2$. The particle–background potential energies are given by

$$E_{pb}(\mathbf{z}) = \sum_{j=1}^{n} V(z_j), \quad E_{pb}^{\pm}(\mathbf{z}) = \sum_{j=1}^{n} V^{\pm}(z_j),$$

respectively, where

$$V(z) := -\frac{n^-}{4\pi^2|\tau|} \int_0^{2\pi} dx' \int_0^{2\pi|\tau|} dy' \, \Phi(z, x' + iy'),$$

$$V^{\pm}(z) := -\frac{n^-}{4\pi^2|\tau|} \int_0^{2\pi} dx' \int_0^{2\pi|\tau|} dy' \, \Phi^{\pm}(z, x' + iy'), \tag{7.12}$$

and the background–background potential energies are given by

$$E_{bb} := -\frac{1}{2} \frac{n^-}{4\pi^2|\tau|} \int_0^{2\pi} dx \int_0^{2\pi|\tau|} dy \, V(x + iy),$$

$$E_{bb}^{\pm} := -\frac{1}{2} \frac{n^-}{4\pi^2|\tau|} \int_0^{2\pi} dx \int_0^{2\pi|\tau|} dy \, V^{\pm}(x + iy), \tag{7.13}$$

respectively. These systems consisting of positive charged particles embedded in negatively charged backgrounds are called the *one-component plasma models* [16]. The probability densities for particle configuration $\mathbf{z} \in (\mathbb{T}^2)^n$ at the *inverse temperature* $\beta := 1/k_B T > 0$ are then given by the *Gibbs distribution* as

$$\mathbf{p}_{plasma}(\mathbf{z}) = \mathbf{p}_{plasma}(\mathbf{z}; n, n^-, \beta)$$
$$:= \frac{e^{-\beta(E_{pp}(\mathbf{z}) + E_{pb}(\mathbf{z}) + E_{bb})}}{Z_{plasma}}, \tag{7.14}$$

$$\mathbf{p}_{plasma}^{\pm}(\mathbf{z}) = \mathbf{p}_{plasma}^{\pm}(\mathbf{z}; n, n^-, \beta)$$
$$:= \frac{e^{-\beta(E_{pp}^{\pm}(\mathbf{z}) + E_{pb}^{\pm}(\mathbf{z}) + E_{bb}^{\pm})}}{Z_{plasma}^{\pm}}, \quad \mathbf{z} \in (\mathbb{T}^2)^n. \tag{7.15}$$

The integrals in (7.12) and (7.13) can be exactly calculated [13, 15, 39]. Then Forrester [13, 15] found a transformation from the neutral ($n^- = n$) plasma model with $\beta = 2$, whose Gibbs distribution is given by $\mathbf{p}_{plasma}(\mathbf{z}; n, n^- = n, \beta = 2)$, to the DPP of type A_{n-1} with the probability density $\mathbf{p}_{\mathbb{T}^2}^{A_{n-1}}$. (See also Lemma 4.1 in [39].) The following equivalences are also proved [39]:

(a) If we set

$$n^- = \frac{\mathcal{N}^{C_n}}{2} = n + 1 \quad \text{and} \quad \beta = 2, \tag{7.16}$$

then the Gibbs distribution (7.15) of the plasma model, $\mathbf{p}_{plasma}^{\pm}(\mathbf{z}; n, n + 1, 2)$, is identified with $\mathbf{p}_{\mathbb{T}^2}^{C_n}(\mathbf{z})$ for the DPP of type C_n.

(b) If we set

$$n^- = \frac{\mathcal{N}^{D_n}}{2} = n - 1 \quad \text{and} \quad \beta = 2, \tag{7.17}$$

then the Gibbs distribution (7.15) of the plasma model, $\mathbf{p}^{\pm}_{\text{plasma}}(\mathbf{z}; n, n - 1, 2)$, is identified with $\mathbf{p}^{D_n}_{\mathbb{T}^2}(\mathbf{z})$ for the DPP of type D_n.

In other words, these one-component plasma models on $\mathbb{T}^2$ are exactly solvable in the sense that all correlation functions are explicitly given by the determinants specified by the correlation kernels as shown in Chap. 6.

It should be noticed that the one-component plasma model with (7.16) is negatively charged by unit, since $n - n^- = n - (n + 1) = -1$, while the plasma model with (7.17) is positively charged by unit, since $n - n^- = n - (n - 1) = 1$. Here we recall that the scaling limits of the DPPs of type C_n and D_n are given by the Ginibre DPPs of types C and D, respectively. As shown by (6.34), the density of points at the origin is zero in type C;

$$\rho^C_{\text{Ginibre}}(0) = 0. \tag{7.18}$$

On the other hand, the density of points at the origin is twice of the bulk density in type D;

$$\rho^D_{\text{Ginibre}}(0) = \frac{1}{\pi} = 2 \lim_{|z| \to \infty} \rho^D_{\text{Ginibre}}(z). \tag{7.19}$$

If we assume that the points of DPPs are positively charged, (7.18) is considered to represent the situation such that a vacancy of a positively charged particle exists at the origin and it makes the system negatively charged by unit, and (7.19) is considered to represent the situation with an addition of a positively charged particle at the origin so that the system is positively charged by unit. In this way, the correspondence of (a) and (b) can be physically interpreted. In addition to the above results, a nontrivial relationship between the exactly solvable one-component plasma models and the *Gaussian free fields* (GFFs) on $\mathbb{T}^2$ [5, 31] has been discussed in [15, 39].

The above results on the DPPs on $\mathbb{T}^2$ of types A_{n-1}, C_n and D_n should be generalized to the DPPs of other four types, $B_n, B_n^\vee, C_n^\vee$, and BC_n. Systematic study of the relationship between DPPs, the one-component plasma models, GFFs, and the *multiple Schramm–Loewner evolution* will be an important future problem [36, 40–42].

Exercise

7.1 Take the limit $t \to \infty, T - t \to \infty$ in (7.4)–(7.9) and derive the temporally homogeneous systems of SDEs. Observe the degeneracy in this limit.

Solutions to Exercises

Chapter 1

1.1 By the definition (1.6),

$$\theta(1/z; p) = \prod_{j=0}^{\infty}(1 - p^j/z)(1 - zp^{j+1})$$

$$= (1 - 1/z)\prod_{j=0}^{\infty}(1 - p^{j+1}/z) \times \frac{\prod_{k=0}^{\infty}(1 - zp^k)}{1 - z} = -\frac{1}{z}\theta(z; p),$$

$$\theta(pz; p) = \prod_{j=0}^{\infty}(1 - zp^{j+1})(1 - p^j/z)$$

$$= \frac{\prod_{j=0}^{\infty}(1 - zp^j)}{1 - z} \times (1 - 1/z)\prod_{\ell=0}^{\infty}(1 - p^{\ell+1}/z) = -\frac{1}{z}\theta(z; p).$$

1.2 By the definition (1.6) of the theta function,

$$\theta(\zeta; p^3)\theta(\zeta p; p^3)\theta(\zeta p^2; p^3)$$

$$= \prod_{\ell=0}^{\infty}\left\{(1 - \zeta p^{3\ell})\left(1 - \frac{p^{3(\ell+1)}}{\zeta}\right)\right\}\left\{(1 - \zeta p^{3\ell+1})\left(1 - \frac{p^{3(\ell+1)}}{\zeta p}\right)\right\}$$

$$\times \left\{(1 - \zeta p^{3\ell+2})\left(1 - \frac{p^{3(\ell+1)}}{\zeta p^2}\right)\right\}$$

© The Author(s), under exclusive license to Springer Nature Singapore Pte Ltd. 2023
M. Katori, *Elliptic Extensions in Statistical and Stochastic Systems*,
SpringerBriefs in Mathematical Physics 47,
https://doi.org/10.1007/978-981-19-9527-9

$$= \prod_{\ell=1}^{\infty} (1 - \zeta p^{3\ell})(1 - \zeta p^{3\ell+1})(1 - \zeta p^{3\ell+2})\left(1 - \frac{\zeta^{3\ell+1}}{\zeta}\right)\left(1 - \frac{\zeta^{3\ell+2}}{\zeta}\right)\left(1 - \frac{\zeta^{3\ell+3}}{\zeta}\right)$$

$$= \prod_{\ell=0}^{\infty} (1 - \zeta p^{\ell})\left(1 - \frac{p^{\ell+1}}{\zeta}\right),$$

which proves (1.20). By the definition of ω_3, $1 + \omega_3 + \omega_3^2 = 0$ and $\omega_3^3 = 1$. Hence, for $\ell \in \mathbb{N}_0$,

$$(1 - \zeta p^{\ell})(1 - \zeta \omega_3 p^{\ell})(1 - \zeta \omega_3^2 p^{\ell})$$
$$= 1 - \zeta(1 + \omega_3 + \omega_3^2)p^{\ell} + \zeta^2 \omega_3(1 + \omega_3 + \omega_3^2)p^{2\ell} - \zeta^3 \omega_3^3 p^{3\ell} = 1 - \zeta^3 p^{3\ell}.$$

Similarly, we see that $\left(1 - \dfrac{p^{\ell+1}}{\zeta}\right)\left(1 - \dfrac{p^{\ell+1}}{\zeta \omega_3}\right)\left(1 - \dfrac{p^{\ell+1}}{\zeta \omega_3^2}\right) = 1 - \dfrac{p^{3(\ell+1)}}{\zeta^3}$ for
$\ell \in \mathbb{N}_0$. Consider the product of these two equalities and take the product over $\ell \in \mathbb{N}_0$. Then (1.21) is obtained.

1.3 Put $x = e^{-2ia}$, $y = e^{-2ib}$, $u = e^{-2ic}$, and $v = e^{-2id}$ in (1.15) and then take the limit $p \to 0$. By applying (1.7), we have, for instance, $\theta(xy; p) \to 1 - xy = 2ie^{-i(a+b)} \sin(a+b)$ as $p \to 0$. Divide both sides of the obtained equation by the common factor $e^{-2i(a+c)}$ to derive (1.22). If we use (1.23), we see that (1.22) is trivial.

Chapter 2

2.1 The following expressions with the nome modular parameter $2\pi i/t$ are obtained: For $t > 0$,

$$\widetilde{p}_{\mathrm{T}}(0, x; t, y) = p(0, x; t, y)\vartheta_0(i\pi(y - x)/t; 2\pi i/t), \qquad x, y \in [0, 2\pi),$$
$$p_{[0,\pi]}^{\mathrm{aa}}(0, x; t, y) = p(0, x; t, y)\{\vartheta_3(i\pi(y - x)/t; 2\pi i/t) - \vartheta_3(i\pi(y + x)/t; 2\pi i/t)\},$$
$$p_{[0,\pi]}^{\mathrm{rr}}(0, x; t, y) = p(0, x; t, y)\{\vartheta_3(i\pi(y - x)/t; 2\pi i/t) + \vartheta_3(i\pi(y + x)/t; 2\pi i/t)\},$$
$$p_{[0,\pi]}^{\mathrm{ar}}(0, x; t, y) = p(0, x; t, y)\{\vartheta_0(i\pi(y - x)/t; 2\pi i/t) - \vartheta_0(i\pi(y + x)/t; 2\pi i/t)\},$$
$$p_{[0,\pi]}^{\mathrm{ra}}(0, x; t, y) = p(0, x; t, y)\{\vartheta_0(i\pi(y - x)/t; 2\pi i/t) + \vartheta_0(i\pi(y + x)/t; 2\pi i/t)\},$$
$$x, y \in [0, \pi].$$

Chapter 3

3.1 Since $(\omega_n\zeta)^n = \zeta^n$, (3.39) is obvious from (3.1). For $j = 1, \ldots, n$, (3.1) gives

$$\psi_j^{A_{n-1}}(p^{1/n}\zeta; p, r) = \zeta^{j-1}p^{(j-1)/n}\theta(p^{j-1}(-1)^{n-1}rp\zeta^n; p^n)$$
$$= p^{(j-1)/n}\zeta^{-1} \times \zeta^j\theta(p^j(-1)^{n-1}r\zeta^n; p^n),$$

which implies (3.40). We also see that

$$\psi_n^{A_{n-1}}(p^{1/n}\zeta; p, r) = \zeta^{n-1}p^{(n-1)/n}\theta(p^{n-1}(-1)^{n-1}rp\zeta^n; p^n)$$
$$= \zeta^{n-1}p^{(n-1)/n}\theta(p^n((-1)^{n-1}r\zeta^n); p^n)$$
$$= -p^{(n-1)/n}\zeta^{-1}(-1)^{n-1}r^{-1}\theta((-1)^{n-1}r\zeta^n; p^n),$$

where (1.10) was used. This implies (3.41).

3.2

(1) Using (1.10), we can show

$$f_1(p\zeta) = R(p\zeta, \zeta_2) := \theta(pr\zeta\zeta_2; p)\zeta_2\theta(p\zeta/\zeta_2; p)$$
$$= \{-\theta(r\zeta\zeta_2; p)/r\zeta\zeta_2\}\zeta_2\{-(\zeta_2/\zeta)\theta(\zeta/\zeta_2 : p)\} = f_1(\zeta)/r\zeta^2.$$

By (1.9),

$$f_2(\zeta) = R(\zeta_1, \zeta) := \theta(r\zeta_1\zeta; p)\zeta\theta(\zeta_1/\zeta; p)$$
$$= \theta(r\zeta_1\zeta; p)\zeta(-\zeta_1/\zeta)\theta(\zeta/\zeta_1; p) = -\theta(r\zeta_1\zeta; p)\zeta_1\theta(\zeta/\zeta_1; p).$$

Then, using (1.10), we can verify that $f_2(p\zeta) = f_2(\zeta)/r\zeta^2$.

(2) By the definition of determinant, $L(\zeta_1, \zeta_1) = 0$ is obvious. We see that

$$L(1/r\zeta_2, \zeta_1) = \det \begin{bmatrix} \theta(-r(1/r\zeta_2)^2; p^2) & \theta(-r\zeta_2^2; p^2) \\ (1/r\zeta_2)\theta(-pr(1/r\zeta_2)^2; p^2) & \zeta_2\theta(-pr\zeta_2^2; p^2) \end{bmatrix}$$
$$= \det \begin{bmatrix} \theta(-1/r\zeta_2^2; p^2) & \theta(-r\zeta_2^2; p^2) \\ (1/r\zeta_2)\theta(-p/r\zeta_2^2; p^2) & \zeta_2\theta(-pr\zeta_2^2; p^2) \end{bmatrix}.$$

By (1.9), $\theta(-1/r\zeta_2^2; p^2) = (1/r\zeta_2^2)\theta(-r\zeta_2^2; p^2)$, and by (1.11), $\theta(-p/r\zeta_2^2; p^2) = \theta(-pr\zeta_2^2; p^2)$. Then $L(1/r\zeta_2, \zeta_2) = 0$. Similarly, we can show that $L(\zeta_1, 1/r\zeta_1) = 0$.

(3) Put $\zeta_1 = 1$ and $\zeta_2 = -1$. Then we obtain

$$L(1, -1) = \begin{bmatrix} \theta(-r; p^2) & \theta(-r; p^2) \\ \theta(-pr; p^2) & -\theta(-pr; p^2) \end{bmatrix} = -2\theta(-r; p^2)\theta(-pr; p^2) = -2\theta(-r; p),$$

where the second equation of (1.12) was used. On the other hand, $R(1, -1) = -\theta(-r; p)\theta(-1; p)$. Hence we have

$$c := L(1, -1)/R(1, -1) = 2/\theta(-1; p)$$
$$= 2/\{(-1; p)_\infty(-p; p)_\infty\} = 1/(-p; p)_\infty^2.$$

Since the first equation of (1.5) gives $(p^2; p^2)_\infty = (p; p)_\infty(-p; p)_\infty$, $c = (p; p)_\infty^2/(p^2; p^2)_\infty^2$ is proved.

3.3 (i) By (3.17) of Lemma 3.2,

$$\psi_j^{R_n}(-\zeta; p) = (-1)^{\alpha_j}\left\{\theta((-1)^{\mathcal{N}}\beta_j(p)\zeta^{\mathcal{N}}; p^{\mathcal{N}}) + (-1)^a\sigma_2\theta((-1)^{\mathcal{N}}\beta_j(p); p^{\mathcal{N}})\right\}.$$

If $R_n = B_n^\vee, C_n, C_n^\vee, D_n$, $\mathcal{N}$ is even by (3.9), and if $R_n = B_n^\vee, C_n, D_n$, $a = 0$ by (3.10). Hence the assertion is proved.
   (ii) By (3.17) of Lemma 3.2,

$$\psi_j^{R_n}(p^{1/2}\zeta; p) = p^{\alpha_j/2}\zeta^{\alpha_j}\theta(\beta_j(p)p^{\mathcal{N}/2}\zeta^{\mathcal{N}}; p^{\mathcal{N}})$$
$$+ \sigma_2 p^{-(\alpha_j-a)/2}\zeta^{-(\alpha_j-a)}\theta(\beta_j(p)p^{-\mathcal{N}/2}\zeta^{-\mathcal{N}}; p^{\mathcal{N}}).$$

By (1.11),

$$\theta(\beta_j(p)p^{\mathcal{N}/2}\zeta^{\mathcal{N}}; p^{\mathcal{N}}) = \theta(p^{\mathcal{N}}(\beta_j(p)p^{-\mathcal{N}/2}\zeta^{\mathcal{N}}); p^{\mathcal{N}}) = \theta(\beta_j(p)^{-1}p^{\mathcal{N}/2}\zeta^{-\mathcal{N}}),$$

and by (1.9), $\theta(\beta_j(p)p^{-\mathcal{N}/2}\zeta^{-\mathcal{N}}; p^{\mathcal{N}}) = -\beta_j(p)p^{-\mathcal{N}/2}\zeta^{\mathcal{N}}\theta(\beta_j(p)^{-1}p^{\mathcal{N}/2}\zeta^{\mathcal{N}}; p^{\mathcal{N}})$. We can see from (3.16) that $\beta_j(p)^{-1}p^{\mathcal{N}/2} = -\sigma_1 p^{-\alpha_j+a/2}$, and then we have the equality

$$\psi_j^{R_n}(p^{1/2}\zeta; p) = \sigma_1\sigma_2 p^{\alpha_j/2}\left\{\zeta^{-(\alpha_j-a+\mathcal{N})}\theta(-\sigma_1 p^{-\alpha_j+a/2}\zeta^{\mathcal{N}}; p^{\mathcal{N}})\right.$$
$$\left. + \sigma_1\sigma_2\zeta^{\alpha_j}\theta(-\sigma_1 p^{-\alpha_j+a/2}\zeta^{\mathcal{N}}; p^{\mathcal{N}})\right\}. \qquad (\text{S.1})$$

If $R_n = C_n, C_n^\vee, D_n$, $\sigma_1 \equiv 1$ and the following equalities are verified:

$$-\sigma_1 p^{-\alpha_j+a/2} = \beta_{n+1-j}, \ \zeta^{-(\alpha_j-a+\mathcal{N})} = \zeta^{-\mathcal{N}/2}\zeta^{\alpha_{n+1-j}}, \ \zeta^{\alpha_j} = \zeta^{-\mathcal{N}/2}\zeta^{-(\alpha_{n+1-j}-a)}.$$

Hence the assertion is proved.

3.4 By (3.17) of Lemma 3.2, $\psi_j^{R_n}(1; p) = \theta(\beta_j(p); p^{\mathcal{N}}) + \sigma_2\theta(\beta_j(p); p^{\mathcal{N}})$. Hence if $\sigma_2 = -1$, then $\psi_j^{R_n}(1; p) = 0$. By (3.12), (3.45) is proved. Following (3.45), Exercise 3.3 proves (3.46), (3.48), and (3.47) for $R_n = C_n, C_n^\vee$. For $R_n = BC_n$, (S.1) gives

$$\psi_j^{BC_n}(p^{1/2}\zeta; p) = -\zeta^{-n}p^{(j-n)/2}\Big\{\zeta^{-j}\theta(-p^{n-j+1/2}\zeta^{2n+1}; p^\mathcal{N})$$
$$-\zeta^j\theta(-p^{n-j+1/2}\zeta^{-2n-1}; p^\mathcal{N})\Big\}.$$

Hence, if we set $\zeta = \pm 1$, we obtain (3.47) and (3.48) for $R_n = BC_n$.

Chapter 4

4.1 By (2.23), (4.24), and (4.25), in this case we have

$$L_j^{D_n} := \sum_{\ell=1}^n a_{j\ell}^{D_n}(p_t)\mathrm{p}_{[0,\pi]}^{\mathrm{rr}}(0, u_\ell; t, x)$$

$$= \frac{2(p_t; p_t)_\infty}{(p_{\mathcal{N}^2 t}; p_{\mathcal{N}^2 t})_\infty} \frac{p_t^{-\gamma_j^2/2}}{\mathcal{N}} \Big\{\sum_{\ell=1}^n \widetilde{c}_\ell \cos(\gamma_j u_\ell)\theta(-p_t^{1/2}e^{i(x-u_\ell)}; p_t)$$

$$+ \sum_{\ell=1}^n \widetilde{c}_\ell \cos(\gamma_j u_\ell)\theta(-p_t^{1/2}e^{i(x+u_\ell)}; p_t)\Big\}$$

$$= \frac{2(p_t; p_t)_\infty}{(p_{\mathcal{N}^2 t}; p_{\mathcal{N}^2 t})_\infty} \frac{p_t^{-\gamma_j^2/2}}{\mathcal{N}} (\widetilde{S}_+ + \widetilde{S}_-),$$

where $\widetilde{S}_\pm := \sum_{\ell=0}^{n-1} \widetilde{c}_{\ell+1} \cos\Big[(j-n)\dfrac{\pi\ell}{n-1}\Big]\theta(-p_t^{1/2}e^{i(x\mp\pi\ell/(n-1))}; p_t)$. Since we have
set $\widetilde{c}_1^{D_n} = \widetilde{c}_n^{D_n} = 1/2 = \widetilde{c}_\ell^{D_n}/2$, $\ell = 2, \ldots, n-1$, if we set $\widetilde{c}_\ell^{D_n} = \widetilde{c}_{-\ell+2}^{D_n}$ for $\ell = -n+2, \ldots, 1$, we can see that

$$\widetilde{S}_+ + \widetilde{S}_- = \frac{1}{2}\Big\{\sum_{\ell=-n+1}^{n-1} \widetilde{c}_{\ell+1} \cos\Big[(j-n)\frac{\pi\ell}{n-1}\Big]\theta(-p_t^{1/2}e^{i(x-\pi\ell/(n-1))}; p_t)$$

$$+ \sum_{\ell=-n+1}^{n-1} \widetilde{c}_{\ell+1} \cos\Big[(j-n)\frac{\pi\ell}{n-1}\Big]\theta(-p_t^{1/2}e^{i(x+\pi\ell/(n-1))}; p_t)\Big\}$$

$$= \frac{1}{2}\Big\{\sum_{\ell=-n+2}^{n-1} \cos\Big[(j-n)\frac{\pi\ell}{n-1}\Big]\theta(-p_t^{1/2}e^{i(x-\pi\ell/(n-1))}; p_t)$$

$$+ \sum_{\ell=-n+2}^{n-1} \cos\Big[(j-n)\frac{\pi\ell}{n-1}\Big]\theta(-p_t^{1/2}e^{i(x+\pi\ell/(n-1))}; p_t)\Big\}.$$

Therefore,

$$
L_j^{D_n} = \frac{(p_t; p_t)_\infty}{(p_{(2n-2)^2t}; p_{(2n-2)^2t})_\infty} \frac{p_t^{-(j-n)^2/2}}{2n-2} \sum_{\ell=-n+2}^{n-1} \cos\left[(j-n)\frac{\pi\ell}{n-1}\right]
$$
$$
\times \left\{ \theta(-p_t^{1/2} e^{i(x-\pi\ell/(n-1))}; p_t) + \theta(-p_t^{1/2} e^{i(x+\pi\ell/(n-1))}; p_t) \right\},
$$

where the ranges of the two summations are both $\{-n+2, \ldots, n-1\}$ consisting of $2n-2 = \mathcal{N}^{D_n}$ elements. Following the calculations similar to those given in (i) and (ii) of the proof for Lemma 4.4, (4.26) is verified for $R_n = D_n$.

Chapter 5

5.1 Using the Laurent expansion (1.13) of the theta function, we have

$$
\int_\mathbb{R} \psi_u^A(e^{ix}; p_t) \overline{\psi_w^A(e^{ix}; p_{T-t})} dx = \frac{2\pi}{(p_t; p_t)_\infty (p_{T-t}; p_{T-t})_\infty}
$$
$$
\times \sum_{\ell\in\mathbb{Z}} \sum_{m\in\mathbb{Z}} p_t^{\binom{\ell}{2}} p_{T-t}^{\binom{m}{2}} p_t^{(u+1/2)\ell} p_{T-t}^{(w+1/2)m} \times \frac{1}{2\pi} \int_{-\infty}^\infty e^{i\{(u-w)+(\ell-m)\}x} dx,
$$

$$
\int_\mathbb{R} \psi_u^R(e^{ix}; p_t) \overline{\psi_w^R(e^{ix}; p_{T-t})} dx = \frac{2\pi}{(p_t; p_t)_\infty (p_{T-t}; p_{T-t})_\infty}
$$
$$
\times \sum_{\ell\in\mathbb{Z}} \sum_{m\in\mathbb{Z}} p_t^{\binom{\ell}{2}} p_{T-t}^{\binom{m}{2}} p_t^{(u+1/2)\ell} p_{T-t}^{(w+1/2)m}
$$
$$
\times \left\{ \frac{1}{2\pi} \int_{-\infty}^\infty e^{i\{(u-w)/2+(\ell-m)\}x} dx \mp \frac{1}{2\pi} \int_{-\infty}^\infty e^{i\{(u+w)/2+(\ell+m)\}x} dx \right\}, \quad (S.2)
$$

where we should take the minus sign for $R = B$ and C and the plus sign for $R = D$ in the last line in (S.2). Since $u, w \in (0, 1)$ and $\ell, m \in \mathbb{Z}$, we see that $u - w \in (-1, 1)$, $(u + w)/2 \in (0, 1)$, $\ell - m \in \mathbb{Z}$ and $\ell + m \in \mathbb{Z}$, and hence

$$
\frac{1}{2\pi} \int_{-\infty}^\infty e^{i\{(u-w)+(\ell-m)\}x} dx = \frac{1}{2\pi} \int_{-\infty}^\infty e^{i\{(u-w)/2+(\ell-m)\}x} dx = \delta_{\ell m} \delta(u - w),
$$

while $\dfrac{1}{2\pi} \displaystyle\int_{-\infty}^\infty e^{i\{(u+w)/2+(\ell+m)\}x} dx = 0$. Therefore, the assertions are proved.

Chapter 6

6.1 The definition (6.18) gives

$$\Psi_{j\pm n}^{A_{n-1}}(z) = \frac{e^{-\pi|\tau|\{(j\pm n)-1\}^2/2}}{\sqrt{C_1(n|\tau|)}} e^{-ny^2/4\pi|\tau|} e^{i\{(j\pm n)-1\}z} \theta(-p^{(j\pm n)-1+n/2}e^{inz}; p^n).$$

By (1.10), $\theta(-p^{j-1+n/2}p^n e^{inz}; p^n) = p^{-(j-1)}p^{-n/2}e^{-inz}\theta(-p^{j-1+n/2}e^{inz}; p^n)$. Since the equality $\theta(-p^{j-1+n/2}e^{inz}; p^n) = \theta(-p^{-j+1+n/2}e^{-inz}; p^n)$ holds by (1.11), (1.10) also proves $\theta(-p^{j-1+n/2}p^{-n}e^{inz}; p^n) = p^{j-1}p^{-n/2}e^{inz}\theta(-p^{j-1+n/2}e^{inz}; p^n)$. Combining them with the equalities

$$e^{-\pi|\tau|\{(j\pm n)-1\}^2/n} = e^{-\pi|\tau|(j-1)^2/n}p^{\pm(j-1)}p^{n/2}, \quad e^{i\{(j\pm n)-1\}z} = e^{i(j-1)z}e^{\pm inz},$$

the equalities (6.38) are proved.

6.2 Notice that $e^{-\pi|\tau|(j-1)^2/n} = e^{-\pi|\tau|j^2/n}p^{-j/n}p^{1/2n}$. We can show that

$$e^{-n(y+2\pi|\tau|/n)^2/4\pi|\tau|} = e^{-ny^2/4\pi|\tau|}e^{-y}p^{1/2n},$$

and hence

$$e^{-\pi|\tau|(j-1)^2/n}e^{-n(y+2\pi|\tau|/n)^2/4\pi|\tau|} = \left\{ e^{-\pi|\tau|j^2/n}e^{-ny^2/4\pi|\tau|} \right\} \times e^{-y}p^{-(j-1)/n}.$$

By (3.40) of Exercise 3.1 and the definition (6.18),

$$\Psi_j^{A_{n-1}}(z+2\pi|\tau|i/n) = e^{-y}e^{-iz}\Psi_{j+1}^{A_{n-1}}(z) = e^{-ix}\Psi_{j+1}^{A_{n-1}}(z), \quad j = 1, \ldots, n-1.$$

When $r = (-1)^n p^{n/2}$, (3.41) of Exercise 3.1 gives

$$\Psi_n^{A_{n-1}}(p^{1/n}\zeta; r, (-1)^n p^{n/2}) = p^{(n-1)/n}\zeta^{-1}p^{n/2}.$$

Since $e^{-\pi|\tau|(n-1)^2/n}e^{-n(y+2\pi|\tau|/n)^2/4\pi|\tau|} = e^{-ny^2/4\pi|\tau|}e^{-y}p^{n/2}p^{-(n-1)/n}$, (6.18) gives

$$\Psi_n^{A_{n-1}}(z+2\pi|\tau|i/n) = e^{-ix}\Psi_1(z).$$

Then (6.28) of Theorem 6.1 implies $K_{\mathbb{T}^2}^{A_{n-1}}(z+2\pi|\tau|i/n, z'+2\pi|\tau|i/n) = K_{\mathbb{T}^2}^{A_{n-1}}(z, z'), z, z' \in \mathbb{T}^2$, and (6.30) of Proposition 6.2 (ii) is proved.

6.3 We can verify the equality

$$e^{-\pi|\tau||J(j)|^2/\mathcal{N}}e^{-\mathcal{N}(y+\pi|\tau|)^2/4\pi|\tau|+a(y+\pi|\tau|)/2}$$

$$= \left\{ e^{-\pi|\tau||J(n+1-j)|^2/\mathcal{N}}e^{-\mathcal{N}y^2/4\pi|\tau|+ay/2} \right\} \times e^{-\mathcal{N}y/2}p^{-\alpha_j/2}$$

for $R_n = C_n$, $C_n^\vee$, and D_n. Combining this with (3.44) of Exercise 3.3, (6.19) gives

$$\Psi_j^{R_n}(z+\pi|\tau|i)=\sigma_2 e^{-i\mathcal{N}z/2}e^{-\mathcal{N}y/2}\Psi_{n+1-j}^{R_n}(z)=\sigma_2 e^{-i\mathcal{N}x/2}\Psi_{n+1-j}^{R_n}(z), \quad j=1,\dots,n,$$

for $R_n = C_n$, $C_n^\vee$, and D_n. By (6.28) of Theorem 6.1, the correlation kernels are invariant for the shift by $\pi|\tau|i$ and (6.32) of Proposition 6.2 (ii) is proved.

6.4 Since $|\tau| \to \infty$ means $p = e^{-2\pi|\tau|} \to 0$, we see that $\theta(-p^{j-1+n/2}e^{inz}; p^n) \to 1$, if $j-1+n/2 > 0$. Hence it is easy to verify that

$$\lim_{\substack{n\to\infty,|\tau|\to\infty, \\ \varrho=\text{const.}}} K_{\mathbb{T}^2}^{A_{n-1}}(z, z') = \frac{\varrho^{1/2}}{2^{1/2}\pi}e^{-\pi\varrho(y^2+y'^2)} \widehat{K}_{\mathbb{T}\times\mathbb{R}}^A((x, y), (x', y')),$$

$(x, y), (x', y') \in \mathbb{T} \times \mathbb{R}$, where

$$\widehat{K}_{\mathbb{T}\times\mathbb{R}}^A((x, y), (x', y')) = \sum_{\ell=-\infty}^{\infty} e^{-\ell^2/2\pi\varrho}e^{i\ell(z-z')}$$

$$= (\widehat{p}; \widehat{p})_\infty \theta(-\widehat{p}^{1/2}e^{i(x-x')-(y-y')}; \widehat{p})$$

with $\widehat{p} := e^{-1/\pi\varrho}$. The reference measure can be chosen as $\lambda_{\mathbb{T}}(dx) \times \lambda_{\mathrm{N}(0,1/4\pi\varrho)}(dy)$, where $\lambda_{\mathbb{T}}(dx) := dx/2\pi$ (the uniform distribution on $\mathbb{T}$) and $\lambda_{\mathrm{N}(0,\sigma^2)}(dy) := e^{-y^2/2\sigma^2}dy/\sqrt{2\pi}\sigma$ (the centered normal distribution with variance σ^2). Notice that the density with respect to $\lambda_{\mathbb{T}}(dx) \times \lambda_{\mathrm{N}(0,1/4\pi\varrho)}(dy)$ is constant, $\widehat{\rho}_{\mathbb{T}\times\mathbb{R}}^A \equiv (\widehat{p}; \widehat{p})_\infty \theta(-\widehat{p}^{1/2}; \widehat{p})$.

Chapter 7

7.1 In the limit $T-t \to \infty$, $p_{\mathcal{N}^{R_n}(T-t)} \to 0$ and we can use (1.7). The SDEs (7.4), (7.7) and (7.8) for the types B_n, $C_n^\vee$ and BC_n are degenerated to

$$dY_j^{B_n}(t) = d\widetilde{B}_j^B(t) + \frac{1}{2}\cot\left(\frac{Y_j^{B_n}(t)}{2}\right)dt$$

$$+ \frac{1}{2}\sum_{1\leq k\leq n, k\neq j}\left[\cot\left\{\frac{1}{2}\left(Y_j^{B_n}(t) - Y_k^{B_n}(t)\right)\right\} + \cot\left\{\frac{1}{2}\left(Y_j^{B_n}(t) + Y_k^{B_n}(t)\right)\right\}\right]dt,$$

$$\text{(S.3)}$$

$j = 1, \ldots, n$, $t \geq 0$, where a reflecting wall is imposed at π. The SDEs (7.5) and (7.6) for the types $B_n^\vee$ and C_n are degenerated to

$$dY_j^{C_n}(t) = d\widetilde{B}_j^C(t) + \cot\left(Y_j^{C_n}(t)\right)dt$$

$$+ \frac{1}{2} \sum_{1 \leq k \leq n, k \neq j} \left[\cot\left\{\frac{1}{2}\left(Y_j^{C_n}(t) - Y_k^{C_n}(t)\right)\right\} + \cot\left\{\frac{1}{2}\left(Y_j^{C_n}(t) + Y_k^{C_n}(t)\right)\right\}\right]dt,$$

$$\text{(S.4)}$$

$j = 1, \ldots, n$, $t \geq 0$. The SDE (7.9) for the type D_n is reduced to

$$dY_j^{D_n}(t) = d\widetilde{B}_j^D(t)$$

$$+ \frac{1}{2} \sum_{1 \leq k \leq n, k \neq j} \left[\cot\left\{\frac{1}{2}\left(Y_j^{D_n}(t) - Y_k^{D_n}(t)\right)\right\} + \cot\left\{\frac{1}{2}\left(Y_j^{D_n}(t) + Y_k^{D_n}(t)\right)\right\}\right]dt,$$

$$\text{(S.5)}$$

$j = 1, \ldots, n$, $t \geq 0$, where reflecting walls are imposed at 0 and π. Here $(\widetilde{B}_j^R(t))_{t \geq 0}$, $j = 1, \ldots, n$, $R = B, C, D$ are mutually independent one-dimensional standard Brownian motions. Notice that the functions appearing in the second terms of the right-hand sides of (S.3) and (S.4) behave as $\cot(y/2) \uparrow \infty$ as $y \downarrow 0$ and $\cot(y/2) \downarrow 0$ as $y \uparrow \pi$, and as $\cot y \uparrow \infty$ as $y \downarrow 0$ or $y \uparrow \pi$. There is no such term in (S.5). They are consistent with the boundary conditions at $y = 0$ and $y = \pi$ given by (4.35). The present degeneracy from the total of seven types to the four types is the same as that shown by (4.46)–(4.49) in Section 4.4.

References

1. Biane, P., Pitman, J., Yor, M.: Probability laws related to the Jacobi theta and Riemann zeta functions, and Brownian excursions. Bull. Amer. Math. Soc. **38**, 435–465 (2001)
2. Borodin, A.: Periodic Schur process and cylindric partitions. Duke Math. J. **140**, 391–468 (2007)
3. Borodin, A.N., Salminen, P.: Handbook of Brownian Motion–Facts and Formulae, 2nd edn. Birkhäuser, Basel (2002)
4. Byun, S.-S., Kang, N.-G., Tak, H.-J.: Conformal field theory for annulus SLE: partition functions and martingale-observables. Ann. Math. Phys. **13**(1) (2023)
5. Cardy, J.: Conformal invariance and statistical mechanics. In: Brézin, E., Zinn-Justin, J. (eds.) Fields, Strings and Critical Phenomena, (Les Houches), pp. 169–246. North-Holland, Amsterdam (1990)
6. Charalambides, C.A.: Discrete q-Distributions. Wiley, Hoboken (2016)
7. Cuenca, C., Gorin, V., Olshanski, G.: The elliptic tail kernel. Int. Math. Res. Not. **19**, 14922–14964 (2021)
8. Date, E., Jimbo, M., Kuniba, A., Miwa, T., Okado, M.: Exactly solvable SOS models, local height probabilities and theta function identities. Nucl. Phys. B **290**(FS20), 231–273 (1987)
9. Demni, N., Mouayn, Z.: Polyanalytic reproducing kernels on the quantized annulus. J. Phys. A: Math. Theor. **54**, 17 (2021)
10. Dyson, F.: Missed opportunity. Bull. Amer. Math. Soc. **78**, 635–652 (1972)
11. Forrester, P.J.: Theta function generalizations of some constant term identities in the theory of random matrices. SIAM J. Math. Anal. **21**, 270–280 (1990)
12. Forrester, P.J.: Exact solution of the lockstep model of vicious walkers. J. Phys. A: Math. Gen. **23**, 1259–1273 (1990)
13. Forrester, P.J.: Exact results for the two-dimensional, two-component plasma at $\Gamma = 2$ in doubly periodic boundary conditions. J. Stat. Phys. **61**, 1141–1160 (1990)
14. Forrester, P.J.: Exact results for vicious walker model of domain walls. J. Phys. A: Math. Gen **24**, 203–218 (1991)
15. Forrester, P.J.: Particles in a magnetic field and plasma analogies: doubly periodic boundary conditions. J. Phys. A: Math. Gen. **39**, 13025–13036 (2006)
16. Forrester, P.J.: Log-Gases and Random Matrices. Princeton University Press, Princeton, NJ (2010)

17. Forrester, P.J.: Meet Andréief, Bordeaux 1886, and Andreev, Kharkov 1882–1883. Random Mat.: Theory Appl. **8**, 1930001 (2019)
18. Frobenius, F. G.: Über die elliptischen Functionen zweiter Art. J. reine angew. Mathematics **93**, 53–68 (1882). https://gdz.sub.uni-goettingen.de/volumes/id/PPN243919689?page=1
19. Frobenius, F. G., Stickelberger, L.: Zur Theorie der elliptischen Functionen. J. reine angew. Mathematics **83**, 175–179 (1877). https://gdz.sub.uni-goettingen.de/volumes/id/PPN243919689?page=1
20. Fulmek, M.: Nonintersecting lattice paths on the cylinder. Séminaire Lotharingien Combin. **52**, B52b (2004)
21. Gasper, G., Rahman, M.: Basic Hypergeometric Series, 2nd edn. Cambridge University Press, Cambridge (2004)
22. Gessel, I., Viennot, G.: Binomial determinants, paths, and hook length formulae. Adv. Math. **58**, 300–321 (1985)
23. Ginibre, J.: Statistical ensembles of complex, quaternion, and real matrices. J. Math. Phys. **6**, 440–449 (1965)
24. Graczyk, P., Małecki, J.: Multidimensional Yamada–Watanabe theorem and its applications to particle systems. J. Math. Phys. **54**, 021503/1–15 (2013)
25. Graczyk, P., Małecki, J.: Strong solutions of non-colliding particle systems. Electron. J. Probab. **19**(119), 1–21 (2014)
26. Graczyk, P., Małecki, J.: On squared Bessel particle systems. Bernoulli **25**(2), 828–847 (2019)
27. Hallnäs, M., Noumi, M., Spiridonov, V. P., Warnaar, S. O.: Special issue on elliptic integrable systems, special functions and quantum field theory. SIGMA **16** (2020). https://www.emis.de/journals/SIGMA/elliptic-integrable-systems.html
28. Hasegawa, K.: Ruijsenaars' commuting difference operators as commuting transfer matrices. Commun. Math. Phys. **187**, 289–325 (1997)
29. Imamura, T., Sasamoto, T.: Fluctuations for stationary q-TASEP. Probab. Theory Relat. Fields **174**, 647–730 (2019)
30. Kajihara, Y., Noumi, M.: Multiple elliptic hypergeometric series. An approach from the Cauchy determinant. Indag. Math. (N.S.) **14**, 395–421 (2003)
31. Kang, N.-G., Makarov, N. G.: Calculus of conformal fields on a compact Riemann surface. arXiv:math-ph/1708.07361
32. Karlin, S., McGregor, J.: Coincidence probabilities. Pacific J. Math. **9**, 1141–1164 (1959)
33. Katori, M.: Determinantal martingales and noncolliding diffusion processes. Stoch. Proc. Appl. **124**, 3724–3768 (2014)
34. Katori, M.: Elliptic determinantal process of type A. Probab. Theory Relat. Fields **162**, 637–677 (2015)
35. Katori, M.: Elliptic Bessel processes and elliptic Dyson models realized as temporally inhomogeneous processes. J. Math. Phys. **57**, 103302/1–32 (2016)
36. Katori, M.: Bessel Processes, Schramm–Loewner Evolution, and the Dyson Model. Springer-Briefs in Mathematical Physics, vol. 11. Springer, Singapore (2016)
37. Katori, M.: Elliptic determinantal processes and elliptic Dyson models. SIGMA **13**(079), 36 (2017)
38. Katori, M.: Macdonald denominators for affine root systems, orthogonal theta functions, and elliptic determinantal point processes. J. Math. Phys. **60**, 013301/1-27 (2019)
39. Katori, M.: Two-dimensional elliptic determinantal point processes and related systems. Commun. Math. Phys. **371**, 1283–1321 (2019)
40. Katori, M., Koshida, S.: Conformal welding problem, flow line problem, and multiple Schramm–Loewner evolution. J. Math. Phys. **61**, 083301/1–25 (2020)
41. Katori, M., Koshida, S.: Gaussian free fields coupled with multiple SLEs driven by stochastic log-gases. Adv. Study Pure Math. **87**, 315–340 (2021)
42. Katori, M., Koshida, S.: Three phases of multiple SLE driven by non-colliding Dyson's Brownian motions. J. Phys. A: Math. Theor. **54**, 19 (2021)
43. Katori, M., Shirai, T.: Partial isometries, duality, and determinantal point processes. Random Mat.: Theory Appl. **11**, 70 (2022)

44. Katori, M., Shirai, T.: Zeros of the i.i.d. Gaussian Laurent series on an annulus: Weighted Szegő kernels and permanental-determinantal point processes. Commun. Math. Phys. **392**, 1099–1151 (2022)

45. Katori, M., Tanemura, H.: Symmetry of matrix-valued stochastic processes and noncolliding diffusion particle systems. J. Math. Phys. **45**, 3058–3085 (2004)

46. Katori, M., Tanemura, H.: Noncolliding Brownian motion and determinantal processes. J. Stat. Phys. **129**, 1233–1277 (2007)

47. Katori, M., Tanemura, H.: Non-equilibrium dynamics of Dyson's model with an infinite number of particles. Commun. Math. Phys. **293**, 469–497 (2010)

48. Konno H.: Elliptic Quantum Groups: Representations and Related Geometry. SpringerBriefs in Mathematical Physics, vol. 37. Springer, Singapore (2020)

49. Koornwinder, T.H.: On the equivalence of two fundamental theta identities. Anal. Appl. (Singap.) **12**, 711–725 (2014)

50. Koshida, S.: Free field theory and observables of periodic Macdonald processes. J. Comb. Theory, Ser. A **182**(105473) 42 (2021)

51. Krattenthaler, C.: Advanced determinant calculus: a complement. Linear Algebra Appl. **411**, 68–166 (2005)

52. Liechty, K., Wang, D.: Nonintersecting Brownian motions on the unit circle. Ann. Probab. **44**, 1134–1211 (2016)

53. Lindström, B.: On the vector representations of induced matroids. Bull. London Math. Soc. **5**, 85–90 (1973)

54. Macdonald, I.G.: Affine root systems and Dedekind's η-function. Invent. Math. **15**, 91–143 (1972)

55. Mehta, M.L.: Random Matrices, 3rd edn. Elsevier, Amsterdam (2004)

56. Nagao, T., Forrester, P.J.: Dynamical correlations for circular ensembles of random matrices. Nucl. Phys. B **660**, 557–578 (2003)

57. Olver, F. W. J., Lozier, D. W., Boisvert, R. F., Clark, C. W. (eds.): NIST Handbook of Mathematical Functions. U.S. Department of Commerce, National Institute of Standards and Technology, Washington, DC, Cambridge University Press, Cambridge (2010). http://dlmf.nist.gov

58. Osada, H.: Dirichlet form approach to infinite-dimensional Wiener processes with singular interactions. Commun. Math. Phys. **176**, 117–131 (1996)

59. Osada, H.: Infinite-dimensional stochastic differential equations related to random matrices. Probab. Theory Relat. Fields **153**, 471–509 (2012)

60. Osada, H.: Interacting Brownian motions in infinite dimensions with logarithmic interaction potentials. Ann. Probab. **41**, 1–49 (2013)

61. Osada, H., Tanemura, H.: Infinite-dimensional stochastic differential equations and tail σ-fields. Probab. Theory Relat. Fields **177**, 1137–1242 (2020)

62. Rains, E.M.: Transformations of elliptic hypergeometric integrals. Ann. Math. **171**, 169–243 (2010)

63. Rosengren, H.: Sums of triangular numbers from the Frobenius determinant. Adv. Math. **208**, 935–961 (2007)

64. Rosengren, H., Schlosser, M.: Elliptic determinant evaluations and the Macdonald identities for affine root systems. Compositio Math. **142**, 937–961 (2006)

65. Ruijsenaars, S.N.M.: Complete integrability of relativistic Calogero-Moser systems and elliptic function identities. Commun. Math. Phys. **110**, 191–213 (1987)

66. Schlosser, M.: Elliptic enumeration of nonintersecting lattice paths. J. Combin. Theory Ser. A **114**, 505–521 (2007)

67. Schlosser, M. J., Spiridonov, V. P., Warnaar, S. O.: Special issue on elliptic hypergeometric functions and their applications. SIGMA **13** (2017), **14** (2018). https://www.emis.de/journals/SIGMA/EHF2017.html

68. Shirai, T., Takahashi, Y.: Fermion process and Fredholm determinant. In: Begehr, H. G. W., Gilbert, R. P., Kajiwara, J. (eds.), Proceedings of the Second ISAAC Congress, vol. 1, pp. 15–23. Kluwer Academic, Dordrecht (2000)

69. Shirai, T., Takahashi, Y.: Random point field associated with certain Fredholm determinants I: fermion, Poisson and boson point processes. J. Funct. Anal. **205**, 414–463 (2003)
70. Shirai, T., Takahashi, Y.: Random point field associated with certain Fredholm determinants II: fermion shifts and their ergodic and Gibbs properties. Ann. Probab. **31**, 1533–1564 (2003)
71. Soshnikov, A.: Determinantal random point fields. Russian Math. Surveys **55**, 923–975 (2000)
72. Spiridonov, V.P.: Theta hypergeometric series. In: Malyshev, V.A., Vershik, A.M. (eds.) Asymptotic Combinatorics with Applications to Mathematical Physics, pp. 307–327. Kluwer Academic, Dordrecht (2002)
73. Spiridonov, V.P.: Introduction to the theory of elliptic hypergeometric integrals. In: Gritsenko, V.A., Spiridonov, V.P. (eds.) Partition Functions and Automorphic Forms, Moscow Lectures, vol. 5, pp. 271–318. Springer, Heidelberg (2020)
74. Tarasov, V., Varchenko, A.: Geometry of q-hypergeometric functions, quantum affine algebras and elliptic quantum groups. Astérisque **246**, 135 (1997)
75. Toh, P.C.: Generalized mth order Jacobi theta functions and the Macdonald identities. Int. J. Number Theory **04**, 461–474 (2008)
76. Tsai, L.-C.: Infinite dimensional stochastic differential equations for Dyson's model. Probab. Theory Relat. Fields **166**, 801–850 (2016)
77. Warnaar, S.O.: Summation and transformation formulas for elliptic hypergeometric series. Constr. Approx. **18**, 479–502 (2002)
78. Whittaker, E.T., Watson, G.N.: A Course of Modern Analysis, 4th edn. Cambridge University Press, Cambridge (1927)

Index

© The Author(s), under exclusive license to Springer Nature Singapore Pte Ltd. 2023
M. Katori, *Elliptic Extensions in Statistical and Stochastic Systems*,
SpringerBriefs in Mathematical Physics 47,
https://doi.org/10.1007/978-981-19-9527-9

Printed in the United States
by Baker & Taylor Publisher Services